Software Architecture for Busy Developers

Talk and act like a software architect in one weekend

Stéphane Eyskens

Packt>

BIRMINGHAM—MUMBAI

Software Architecture for Busy Developers

Group Product Manager: Aaron Lazar

Publishing Product Manager: Kunal Chaudhari

Senior Editor: Ruvika Rao

Content Development Editor: Kinnari Chohan

Technical Editor: Pradeep Sahu

Copy Editor: Safis Editing

Project Coordinator: Deeksha Thakkar

Proofreader: Safis Editing

Indexer: Vinayak Purushotham

Production Designer: Aparna Bhagat

First published: August 2021

Production reference: 1170821

Published by Packt Publishing Ltd.
Livery Place
35 Livery Street
Birmingham
B3 2PB, UK.

ISBN 978-1-80107-159-8

www.packt.com

Contributors

About the author

Stéphane Eyskens has a developer background and became a solution architect about a decade ago. As a cloud subject matter expert, he contributed to many digital transformation programs, helping organizations get better results out of their cloud investments. As an MVP, he is an active contributor to the Microsoft Tech Community and has worked on multiple open source projects available on GitHub. Stéphane is also a Pluralsight assessment author as well as the author of multiple books and online recordings.

About the reviewers

Sagar Sharma is a Microsoft certified Azure architect and a solution architect with 12 years of professional experience, currently based in the Netherlands. He is also a blogger and frequent public speaker.

Sagar helps customers from various domains by designing solution architecture based on ArchiMate and Open Group standards. He has designed and implemented enterprise software solutions in the fields of the cloud, app modernization, data, integration, and IoT. He also works on defining cloud strategies and enjoys training people on the Azure platform.

He is an Indian by nationality and married to Pragya. They have two lovely kids, Dhruv and Dhwani. He is a movie buff, loves traveling, and is passionate about cooking.

Email: imsharmasagar@outlook.com

Thomas Browet has a passion for IT that started 20 years ago when he discovered Linux while he was still a teenager. He was immediately hooked. After graduating, he started his career writing software for local businesses. He had to develop/plan/configure everything himself. After a few years, he moved to the enterprise world, where everything was more complex, but the core was the same.

Recognized as a thought leader in automation and software development, he leverages his one-man-shop experience to ease collaboration across IT organizations' silos. After a decade of software development, he now works in Brussels as a freelance solution architect where he architects/automates software for the cloud. A father of two, he enjoys rock climbing.

Table of Contents

3

Understanding ATAM and the Software Quality Attributes

Section 3: Software Design Patterns and Architecture Models

4

Reviewing the Historical Architecture Styles

5

Design Patterns and Clean Architecture

Section 4: Impact of the Cloud on Software Architecture Practices

6

Impact of the Cloud on the Software Architecture Practice

Section 5: Architectural Trends and Summary

7

Trendy Architectures and Global Summary

Other Books You May Enjoy

Index

Preface

Software architecture is a broad topic and there is not one single definition of it. In this book, I will try to share my experience in the field, with various customers within different industries. I will take a pragmatic approach to fulfill this book's tagline: Talk and act like a software architect in a weekend. That's all it takes to grasp most concepts and to get started. Of course, you will need to look more deeply into some topics on your own, and this might take a little longer than a weekend. The book will take you on a software architecture journey as practiced in the real world: no fluff and actionable reading.

Who this book is for

This book is for developers who wish to move up the organizational ladder to become software architects. It will help them understand the broader application landscape and how large enterprises deal with software architecture practices. Prior knowledge of software development is required to get the most out of this book.

What this book covers

Chapter 1, Introducing Software Architecture, introduces software architecture and how it is reflected in the real world.

Chapter 2, Exploring Architecture Frameworks and Methodologies, analyzes further the frameworks that we briefly introduced in the previous chapter, through actionable examples.

Chapter 3, Understanding ATAM and the Software Quality Attributes, introduces ATAM, a methodology that you can use to find the most suitable architecture for software.

Chapter 4, Reviewing the Historical Architectural Styles, revisits the history of monoliths, service-oriented architecture, and microservices. It's important to know what has happened over the past decade because architects often need to refactor/migrate existing solutions.

Chapter 5, Design Patterns and Clean Architecture, explores software development patterns and the latest trends with regard to structuring and designing applications.

Chapter 6, *Impact of the Cloud on the Software Architecture Practices*, walks you through the most important aspects to grasp when dealing with the cloud and cloud-native applications.

Chapter 7, *Architectural Trends and Global Summary*, focuses on the most in-demand software architectures and summarizes what we have learned in this book.

To get the most out of this book

Prior knowledge of software development is expected to have the best reading experience. Since the software architecture topic itself is technology-agnostic, you do not need language-specific skills nor language-specific tools. However, for the sake of demonstration, some examples are based on .NET, Azure, and Kubernetes. If you want to reproduce them in your own environment, you will need the following:

Software	Operating system requirements
.NET Core or .NET 5	
An Azure subscription	Any operating system
Docker and Kubernetes	

Rest assured that you will be able to fully grasp the concepts discussed in this book even if you decide not to replay the examples by yourself.

If you are using the digital version of this book, we advise you to type the code yourself or access the code from the book's GitHub repository (a link is available in the next section). Doing so will help you avoid any potential errors related to the copying and pasting of code.

Download the example code files

You can download the example code files for this book from GitHub at `https://github.com/PacktPublishing/Software-Architecture-for-Busy-Developers`. If there's an update to the code, it will be updated in the GitHub repository.

We also have other code bundles from our rich catalog of books and videos available at `https://github.com/PacktPublishing/`. Check them out!

Download the color images

We also provide a PDF file that has color images of the screenshots and diagrams used in this book. You can download it here: `https://static.packt-cdn.com/downloads/9781801071598_ColorImages.pdf`.

Conventions used

There are a number of text conventions used throughout this book.

`Code in text`: Indicates code words in text, database table names, folder names, filenames, file extensions, pathnames, dummy URLs, user input, and Twitter handles. Here is an example: "Let's say that we have a `Rectangle` base class with two separate `SetWitdh` and `SetHeight` methods."

A block of code is set as follows:

```
Rectangle rect = new Square();
rect.setWidth(10);
rect.setHeight(5);
Assert.Equal(50, CalculateArea(rect));
```

Bold: Indicates a new term, an important word, or words that you see onscreen. For instance, words in menus or dialog boxes appear in **bold**. Here is an example: "I defined the **MessageBroker** as an ABB, and the three rectangles on the right are solutions that fulfil this need."

> Tips or important notes
> Appear like this.

Get in touch

Feedback from our readers is always welcome.

General feedback: If you have questions about any aspect of this book, email us at `customercare@packtpub.com` and mention the book title in the subject of your message.

Errata: Although we have taken every care to ensure the accuracy of our content, mistakes do happen. If you have found a mistake in this book, we would be grateful if you would report this to us. Please visit `www.packtpub.com/support/errata` and fill in the form.

Piracy: If you come across any illegal copies of our works in any form on the internet, we would be grateful if you would provide us with the location address or website name. Please contact us at `copyright@packt.com` with a link to the material.

If you are interested in becoming an author: If there is a topic that you have expertise in and you are interested in either writing or contributing to a book, please visit `authors.packtpub.com`.

Share Your Thoughts

Once you've read *Software Architecture for Busy Developers*, we'd love to hear your thoughts! Scan the QR code below to go straight to the Amazon review page for this book and share your feedback.

https://packt.link/r/1801071594

Your review is important to us and the tech community and will help us make sure we're delivering excellent quality content.

Section 1: Introduction

In this part, I'll introduce software architecture, what we are talking about and how it is reflected in the real world. I'll also explain what is in scope for the book and what's not.

This section comprises the following chapter:

- *Chapter 1, Introducing Software Architecture*

1
Introducing Software Architecture

In this chapter, I will introduce the subject of software architecture. My purpose is to help you understand my vision of software architecture and how I will tackle this topic throughout the book.

More specifically, this chapter covers the following topics:

- Software architecture in a nutshell
- A software architect's duties
- Introducing the different architecture disciplines
- Positioning software architecture within the global architecture landscape

By the end of the chapter, you should have a better grip on software architecture and a better idea of how this book will walk you through your software architecture journey.

Software architecture in a nutshell

However rich the literature is on the topic, it's not so easy to find a common definition of software architecture. We as architects like to decouple things, so let's decouple the words **software** and **architecture**. Starting with **software**, we can give this broad definition: *computer programs*. Our second word, **architecture**, can be defined as *the art of designing buildings, houses, and the like*. If we extrapolate a bit, we could define software architecture as *the art of designing computer programs*. This definition sounds very broad.

Rest assured, we can evacuate **hardware** from the equation because it represents the machines themselves. Phew—this should make our task easier, although we are left with everything that runs on a piece of hardware, which remains extremely vast.

Searching for software architecture on Google reveals about 262,000 results, which is very impressive. So many results probably mean a lot of different definitions and a lack of a common understanding of what software architecture is all about. The same query on Google Images does not bring up a single diagram that could help up figure out what software architecture is.

Given the diversity of definitions, I will provide my own, although I don't claim to have the absolute truth. I will start by tying software architecture to the **Architecture Tradeoff Analysis Method** (**ATAM**), which we will see in *Chapter 3, Understanding ATAM and the Software Quality Attributes*. In a nutshell, ATAM consists of evaluating different quality attributes—such as performance, availability, reliability, and so on—of a piece of software. Some of these attributes are code-related, whereas some are infrastructure- or security-related.

Although there is no single definition of software architecture, one thing is absolutely certain: *a software architect should be acquainted with ATAM*. Another thing that appears as an emerging consensus is that the role of a software architect is tightly coupled with the actual development of an application, and I share this viewpoint. This is how *Wikipedia* (`https://en.wikipedia.org/wiki/Software_architect`) defines software architecture, but for me, software architecture goes far beyond mere coding, and that is what you will find out while reading this book. Let's now look at what a software architect's job description might look like.

A software architect's duties

Sometimes, a good job description helps understand the tasks and duties pertaining to a given position. Here again, looking for such descriptions on the web gives many different results, but this is what I think are the responsibilities of a software architect:

- **Addressing both functional requirements (FRs) and non-functional requirements (NFRs)**: As you know, FRs are the primary trigger to design a solution. Whether the solution/service is designed for the business or for the **Information Technology** (**IT**) department itself, you make your business case and then start the design work. NFRs (availability, security, **disaster recovery** (**DR**), and so on) are not always expressed but are also particularly important and are often the most challenging part.

  This book will help you address these requirements in a structured way. Addressing both FRs and NFRs is also the duty of a solution architect, so it can be debated whether this should fall under the duties of a software architect or not. I think that a *good* software architect is able to address FRs. An *excellent* software architect can address both while addressing FRs in a more in-depth way than a solution architect.

- **Providing technical standards, coding guidelines, and design patterns to developers**: Functional features are an integral part of the code base. It's a no-brainer that good design patterns usually improve the resulting quality of a solution. As a software architect, you must understand them and you must be able to sense which pattern is valuable in your own context. We will dive into design patterns in *Chapter 5, Design Patterns and Clean Architecture*, as well as in *Chapter 6, Impact of the Cloud on the Software Architecture Practice*, where we will explore cloud and cloud-native patterns.

- **Interacting with stakeholders to ensure developed applications land smoothly in the company's landscape**: A successful software architect understands multilateral concerns. They are able to interact with security, infrastructure, solution, and enterprise architects, as well as with developers.

  This book will give you the essential keys to achieve **T-shaped skills**, which means being an expert in a given field (the base of the *T*) but also knowledgeable across disciplines (the top of the *T*). To reinforce your *T* top bar, we will explore some fundamentals of **enterprise architecture** (**EA**), which is a common practice in large organizations, and we will touch on some infrastructure and security typical frameworks. However, I have to manage your expectations here, as the book will only *introduce* these topics.

- **Performing an active role in the solution delivery process**: A close follow-up with the development teams and a good understanding of waterfall and agile methodologies will undoubtedly lead to a smoother delivery.

- **Proactive technology watching to identify new trends and paradigm shifts**: Technology is an enabler. Most digital native companies managed to leverage technology wisely and they became Netflix, Facebook, and the like. A good software architect must permanently keep an eye on the ever-evolving technology landscape.

More importantly, a good software architect must exercise good judgment. They should not blindly apply framework *x* or *y*, nor pattern *a* or *b*. They must contextualize and apply their skills wisely. Let's now discover the various architecture disciplines.

Introducing the different architecture disciplines

There are so many types of architects that we can quickly get lost and wonder who does what in an enterprise. Because this understanding is an essential asset, let's start by reviewing the different disciplines, and I will position software architecture in the mix in the next section. The following diagram shows some of the most common architecture disciplines and their main duties:

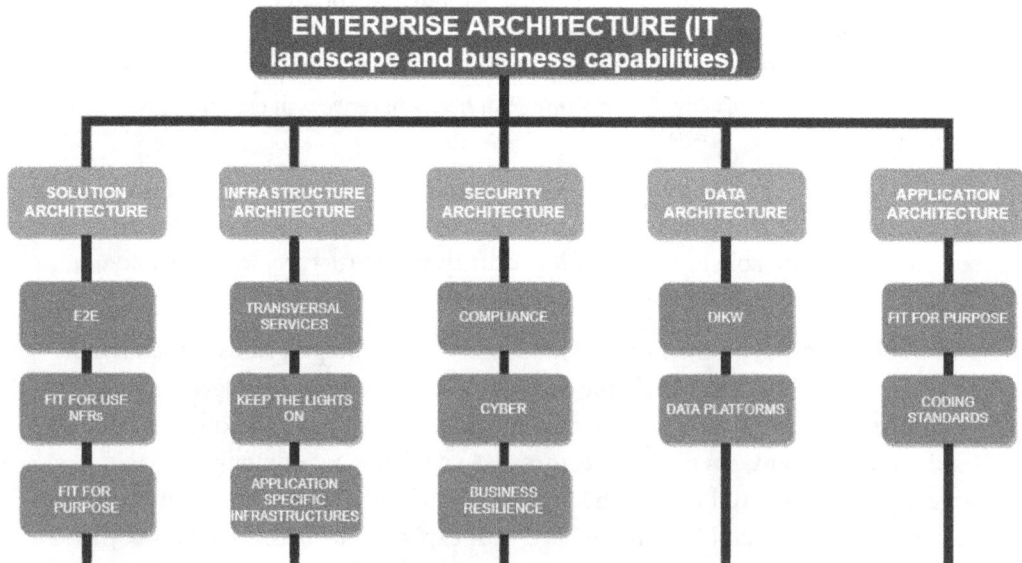

Figure 1.1 – Architecture disciplines: main duties

From top to bottom of the preceding diagram, you can find the main duties by order of priority. Not every discipline is represented, but the main ones are. A noticeable exception, however, is cloud architecture, which we will talk about later. We will discuss some of the related frameworks in our next chapter. Let's now focus on the scope of each discipline.

EA

Enterprise architects oversee the IT and business strategies, and they make sure every IT initiative is in line with the enterprise business goals. They directly report to IT leadership and are sometimes scattered across business lines. They are also the guardians of building coherent and consistent overall IT landscapes for their respective companies.

Most of the time, enterprise architects have a holistic view of the IT landscape and are not concerned with technicalities. Their primary focus is to identify and design business capabilities. They are helped by business architects, who are usually also a part of the EA function. Their role consists of defining the strategic orientations and making sure their standards percolate across teams. They usually work with **The Open Group Architecture Framework (TOGAF)** to define the processes and with ArchiMate to build visual models of the different domains. In some organizations, the EA function can assign a dedicated enterprise architect for large projects or programs.

Solution architecture (SolAr)

Solution architects help different teams to build solutions. They have so-called *T-shaped* skills because they oversee the design of a solution **end to end (E2E)**. They mostly focus on NFRs to ensure a solution is fit for use. As with software architects, they also look at FRs (ensuring that they are fit for purpose), but they are not involved in the actual development of the features.

Infrastructure architecture (IA)

Infrastructure architects focus on building and operating specific application infrastructures and platforms that are shared across workloads. One of their duties is to keep the lights on, to ensure commodity services such as mail systems and workplace-related services are up and running. Infrastructure is organized around **IT Service Management (ITSM)**, which most of the time is based on the **IT Infrastructure Library (ITIL)**. The infrastructure department also provides a service-desk function. Many organizations have started to move (or have moved already) to ServiceNow, a more modern way to handle ITSM.

Security architecture (SA)

In this hyper-connected world, the importance of cybersecurity has grown a lot. SA deals with regulatory or in-house compliance requirements. In these modern times, more and more workloads are deployed to the cloud, which often emphasizes security concerns because many organizations are still in the middle of their cloud journey, or on the verge of starting it.

The security field is split into different sub-disciplines such as **security operations centers (SOCs)**, the management of specialized security hardware and software, **Identity and Access Management (IAM)**, and overall security governance, also known as the **security posture**. In medium-to-large organizations, you can find blue (defend) and red (attack) teams, composed of technical security experts. Together with the SOC, they evaluate the robustness of a business's overall resilience. The SA practice is usually managed by a **chief information security officer (CISO)**, although the role is sometimes also carried out by a **chief information officer (CIO)**.

Depending on the industry you are in, security is typically one of the lesser-known NFRs, not well understood by business, which leads to a complete lack of business requirements in that matter. Security-awareness programs are often required to alert businesses about the importance of security. In a nutshell, the way the security practice is conducted heavily depends on the culture, the risk appetite, and the organization's DNA.

Data architecture (DA)

Data architects oversee the entire data landscape. They mostly focus on designing data platforms for storage, insights, and advanced analytics. They deal with data modeling, data quality, and **business intelligence (BI)**, which consists of extracting valuable insights from the data to realize substantial business benefits. A well-organized data architecture should ultimately deliver the **data-information-knowledge-wisdom (DIKW)** pyramid, as shown in the following diagram:

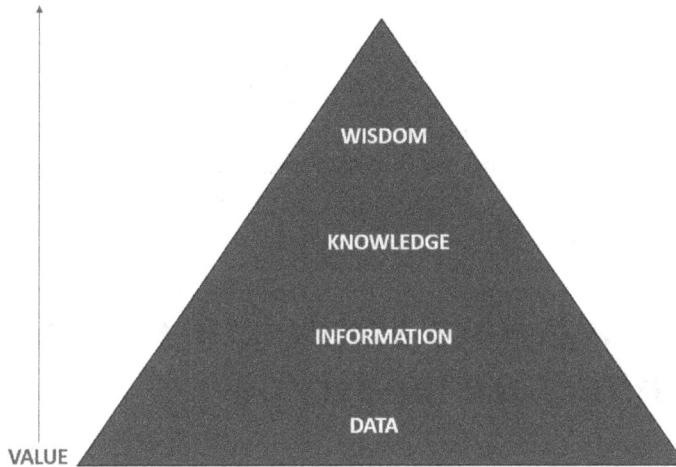

Figure 1.2 – DIKW pyramid

Organizations have a lot of data, from which they try to extract valuable information and knowledge and gain wisdom over time. The more you climb the pyramid, the more value there is. Consider the following scenario to understand the DIKW pyramid:

Data	31, 3, 3000 31, 3, 3100 31, 3, 3500
Information	Day: 31, Month: March, Concurrent Users: 3000
Knowledge	3000 concurrent users visited our web site on March 31. We know that this is way above our daily average, which is about 650 concurrent users. March 31 is always a busy day, year after year.
Wisdom	We'll make sure to restock warehouses before March 31.

Figure 1.3 – DIKW pyramid example

This shows us that, among other things, the work of a data architect is to help organizations learn from their data.

Application architecture (AA)

Application architects focus on building features that are requested by the business. They make sure the developed application is *fit for purpose*. They enforce coding patterns and guidelines to make maintainable and readable applications. Their primary concern is to integrate with the various frameworks and ecosystems. Their role resembles the software architect one but is, in my opinion, more limited. Let's now position software architecture within the various disciplines.

Positioning software architecture within the global architecture landscape

Given the introduction outlined in the previous sections, I will position software architecture very closely to the actual development of a solution, but I will not limit it to only that.

The following screenshot shows how I position the software architecture practice:

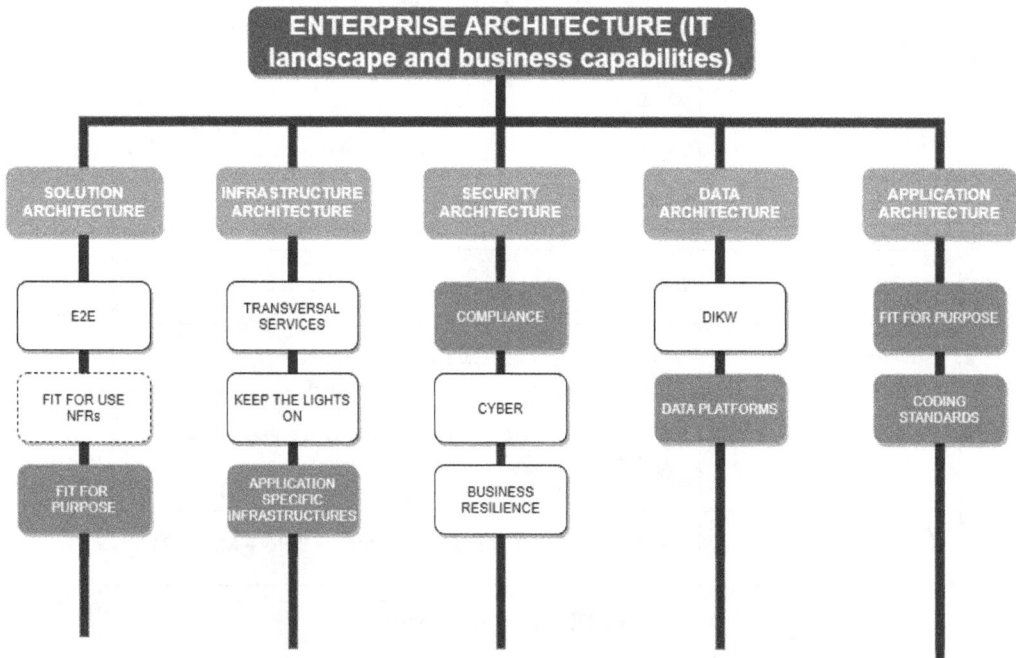

Figure 1.4 – Software architecture within the architecture landscape

Software architects should ideally be knowledgeable in all the topics listed in the shapes with a dark background in *Figure 1.4*. As stated before, software architects will be less focused on NFRs (the shape with dotted lines in *Figure 1.4*) than a solution architect but should still know the basics. The same consideration applies to an EA practice. Let's now recap on this first chapter.

Summary

In this chapter, I set the scene by explaining my understanding of software architecture, which is based on real-world experience within different companies and industries. We reviewed the different architecture disciplines when positioning software architecture, and I wanted you to realize that you need to know a little bit of all the other disciplines to be a successful software architect. I hope that you understand the value proposition of this book and that you are ready to embark on this software architecture adventure.

In the next chapter, I will slightly touch on some of the typical frameworks used in the different disciplines. This will help you speak the vocabulary of your stakeholders to become an even better software architect.

Section 2: The Broader Architecture Landscape

In this part, we will focus on the broader architecture landscape and make sure you understand that software architecture is only a subset of it. As part of its activities, the software architect will interact in one way or another with many other types of architects. Each stakeholder has its own vocabulary and its own specific set of frameworks. I want to ensure that you are adequately equipped when interacting with your peers and stakeholders. I also want to introduce a generic methodology (ATAM) that is heavily used in the enterprise world and across different industries.

This section comprises the following chapters:

- *Chapter 2, Exploring Architecture Frameworks and Methodologies*
- *Chapter 3, Understanding ATAM and the Software Quality Attributes*

2
Exploring Architecture Frameworks and Methodologies

In this chapter, we will explore the most widespread frameworks that you will typically encounter in various organizations. Be aware that I only introduce them because most of these frameworks deserve a dedicated book. I will try to highlight the essential parts and the mindset behind each framework.

More specifically, this chapter covers the following topics:

- Introducing frameworks and methodologies
- Delving into **The Open Group Architecture Framework (TOGAF)**, ArchiMate, and related tools—**enterprise architecture (EA)**
- Introducing security frameworks
- The **Information Technology (IT) Infrastructure Library (ITIL)** in a nutshell

By the end of the chapter, you should be better equipped to interact with stakeholders who contribute to the broader IT landscape. Understanding their concerns is key to growing as an architect. I encourage you to go the extra mile on your own, to brush up your skills in some of the matters we will touch on in this chapter.

Introducing frameworks and methodologies

There are many frameworks, standards, and architecture tools in the IT industry. Frameworks are essentially a set of best practices that should be inspirational for an enterprise's users. They differ from standards in that they are not prescriptive. Conversely, standards *are prescriptive*, in that you must adhere to all their rules to get certified. The following screenshot shows some of the recurring frameworks, standards, and tools, as per the architecture discipline:

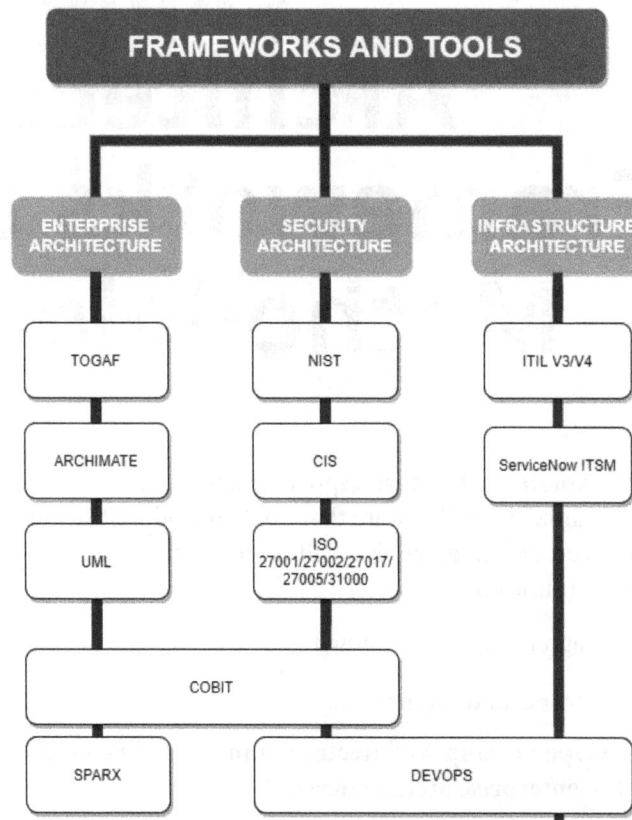

Figure 2.1 – Frameworks and tools

Starting from the left, we have the EA practice, which is mostly conducted using TOGAF and the **ArchiMate** modeling language. **Sparx Systems Enterprise Architect** is a widespread tool suite that helps you build both ArchiMate viewpoints and **Unified Modeling Language** (**UML**) diagrams. ArchiMate is used to draw high-level types of viewpoints, while UML can be used to draw both high- and low-level types of views. If you stick to ArchiMate and have low requirements, you may consider using **Archi** (`https://www.archimatetool.com/`), a free open source software.

Control Objectives for Information and Related Technologies (**COBIT**) can be used to supplement the other frameworks, depending on whether the EA function encompasses the governance body or not. COBIT is mostly used to establish and enforce proper governance within an organization. COBIT also has a specific security-related counterpart—namely, **COBIT for Risk**.

Talking of security, the **National Institute of Standards and Technology** (**NIST**) and **Center for Internet Security** (**CIS**) are well known and followed by security architects. There are a plethora of security-related **International Organization for Standardization** (**ISO**)/**International Electrotechnical Commission** (**IEC**) standards for which you can get certified. From an infrastructure perspective, the historical leader has always been ITIL, which can be de facto implemented by leveraging the ServiceNow **IT Service Management** (**ITSM**) platform. Let's now have a closer look at EA.

Delving into TOGAF, ArchiMate, and related tools – EA

As explained in the first chapter, the main purpose of the EA function is to connect the IT department to different business lines and to make sure that IT investments and initiatives ultimately create business value. Most EA organizations are based on TOGAF.

Introducing TOGAF's ADM

The purpose of the **Architecture Development Method (ADM)** is to manage the development life cycle of the EA practice. The following screenshot shows the ADM proposed by TOGAF:

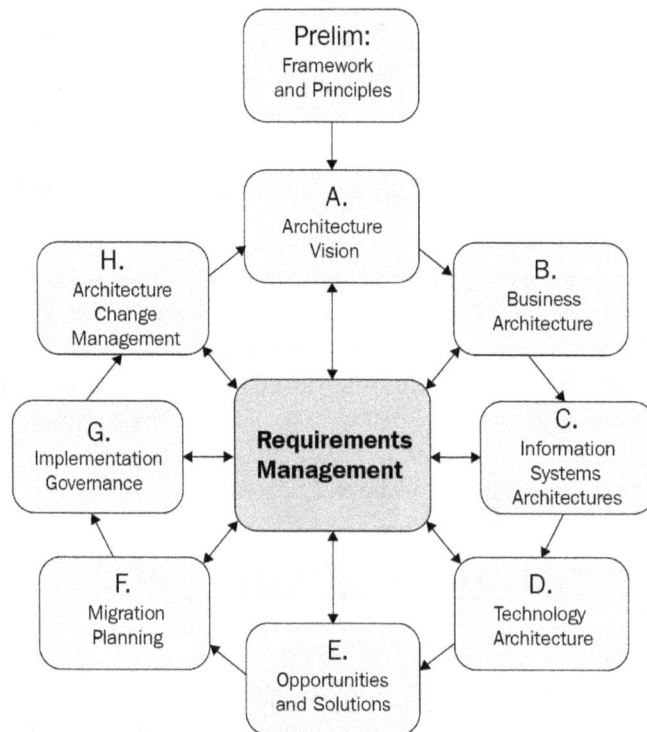

Figure 2.2 – TOGAF ADM

You can follow this to cover the entire spectrum of EA in a series of logical steps. However, because TOGAF is only a framework, you can perfectly decide to work only with some of the steps of that development cycle. The extent to which you apply TOGAF will be better understood once you have completed the preliminary steps and the architecture vision (Step **A** in *Figure 2.2*). If there were only one required step, it would be that step, because it will help you shape your EA practice and decide which other steps you will consider using. However, because this book is not about EA, I will only explain a few important concepts that you are more likely to encounter in your day-to-day work as a software architect. Let's start with the building blocks.

Building blocks

A building block is a business or IT component that can be assembled with other blocks to deliver solutions. You are likely to encounter **architecture building blocks (ABBs)** and **solution building blocks (SBBs)**.

As per the TOGAF specification, ABBs and SBBs are technology-aware, especially when they are used to describe the application landscape. The following screenshot shows an example of ABB and SBB modeling:

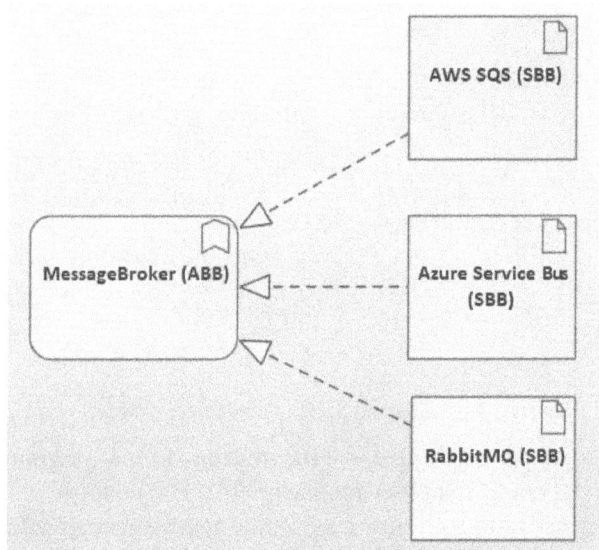

Figure 2.3 – ABB and SBB

An ABB is agnostic to specific implementations. SBBs correspond to the implementation of an ABB, but they represent the actual solution(s). In *Figure 2.3*, I defined **MessageBroker** as an ABB, and the three rectangles on the right are solutions that fulfill this need. When used in a business context, an ABB depicts a business function and an SBB designates one or more solutions that serve this business function.

Our purpose is not to list all possible SBBs but what has been chosen and validated by the enterprise, unlike those shown in the following screenshot:

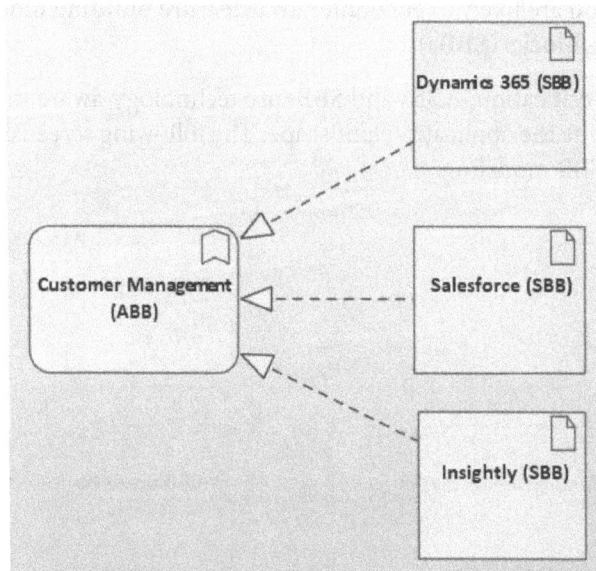

Figure 2.4 – ABBs and SBBs in a business context

I simply did not want to be partial or give you the impression that you should go for *x* or *y* to fulfill such a business need (customer management). The purpose of building blocks is to represent both business and technical needs and their associated solutions. They will help you make consistent choices and prevent acquiring (or building) redundant solutions to serve the same needs.

All the ArchiMate diagrams and viewpoints are high-level and can be presented to any stakeholder (management, IT, and so on). Let's now explore the notion of architecture patterns.

Architecture patterns

Architecture patterns are still high-level but already lower-level diagrams and are closer to the technical implementation of a solution. They are called *patterns* because they can be reused across applications and solutions. They have proven to be valuable on several occasions. As a software architect, you might have to model a few patterns using ArchiMate or UML. The following screenshot shows an example of a microservice pattern:

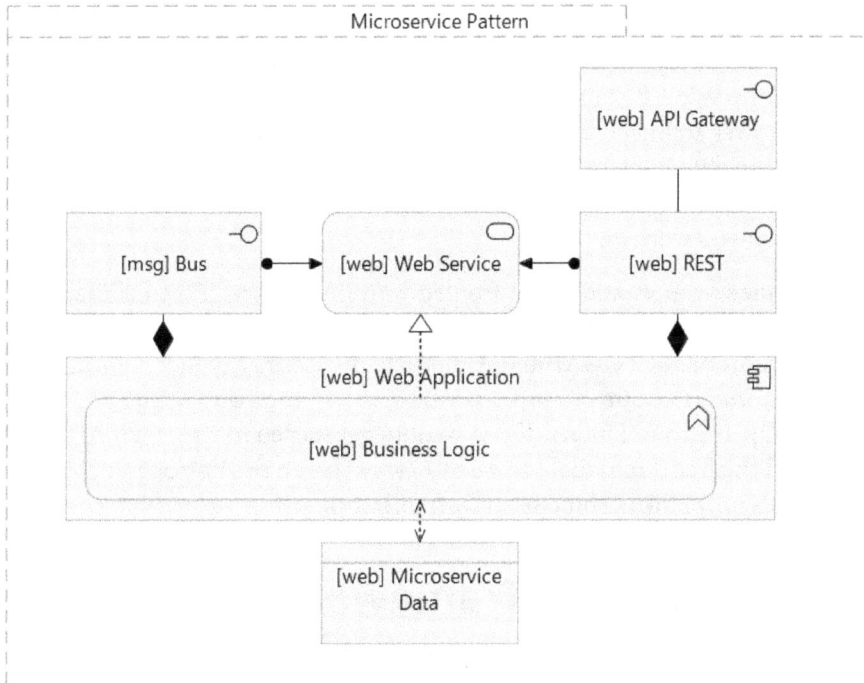

Figure 2.5 – Microservice pattern

ArchiMate has a ton of symbols that represent an object type and the relationships between components. It is far beyond the scope of this book to explain all of them. In a nutshell, *Figure 2.5* shows that the microservice pattern is made of a web application component that encompasses the business logic (application function). That web application is itself composed of web service, **REpresentational State Transfer** (**REST**), and messaging interfaces. ArchiMate interface objects are easily recognizable thanks to their lollipop type of bar. The web application talks to a microservice-specific data store. The REST interface is itself associated with an **application programming interface** (**API**) gateway for client consumption.

Here again, you can assemble *validated* patterns to build a new solution, which will ultimately help you gain time. By validated patterns, I mean patterns that have been assessed from a **non-functional requirements** (**NFR**) perspective (security, scalability, and so on). If you want to build yet lower-level diagrams such as class diagrams, you should consider UML.

EA wrap-up

The EA practice is not always well understood by other stakeholders. It is often perceived as bringing low-to-no value, by both businesses and other IT teams. EA architects are sometimes ivory-tower architects, thinking a lot but not contributing effectively (or efficiently) to the concrete implementation of a solution. To prevent such a situation, you must work hard on the architecture vision piece and make sure it is suited for the size and type of organization you work in.

As a software architect, you should be confronted with EA artifacts if the EA practice is properly conducted because a well-driven practice must percolate across layers. If you want to read more about TOGAF and ArchiMate, I encourage you to look at the *ArchiSurance Case Study* (`https://publications.opengroup.org/y163`), published by the Open Group Library. If you want to get started for free with ArchiMate, you can download the free **Archi** tool (`https://www.archimatetool.com/download/`). Let's now look at some security frameworks.

Introducing security frameworks

Before exploring security frameworks, let me describe the typical duties of a security architect. One of the best ways to identify them is to look at **Certified Information Systems Security Professional (CISSP)** certification, which is the most wanted certification for security professionals. The CISSP exam covers the following topics:

- Security and Risk Management
- Asset Security
- Security Architecture and Engineering
- Communication and Network Security
- **Identity and Access Management (IAM)**
- Security Assessment and Testing
- **Security Operations Center (SOC)**
- Software Development Security

The list is composed of both IT and technical security topics. I don't know about you, but as a cloud architect I am heavily exposed to security demands, and I consider that being able to talk the language of a security architect is a key asset to overcome some hurdles.

A properly driven security organization revolves around the risk management function. Being able to assess the risk related to a business asset will allow you to understand which security controls and processes should be applied accordingly. It is also important to understand the risk appetite of an organization. Some are risk-averse (typically the banking sector) while some are risk-friendly (typically start-ups). Many industries are also subject to specific regulations (compliance). Software architects must be able to evaluate the security culture of the organization they work for, to tackle security more efficiently.

The shift-left paradigm, which means integrating security from the ground up, is not yet mainstream. As a software architect, you will primarily be confronted with software development security. Instead of trying to become a security expert yourself, you should advocate for a strong **continuous integration/continuous delivery (CI/CD)** factory that encompasses static-code security scanners and uses them as quality gates to promote a developed component from a development environment to another environment that is closer to production.

This first step can already reassure a security organization that you are seriously considering security. The factory is part of the **development-operations (DevOps)** toolchain, which is sometimes even called **development-security-operations (DevSecOps)** or **GitOps**, or even **GitSecOps**. When adding *Sec* in the middle, you clearly emphasize the security concern. Make sure you get familiar with this way of working if you are not already aware of it.

Because this book is not about security, I will not cover all the topics. However, I promised I would tackle a few frameworks, so let me start with COBIT 5 for risk.

COBIT for risk

COBIT is maintained by the **Information Systems Audit and Control Association (ISACA)** organization (`https://www.isaca.org/`). What I like about the COBIT framework in general is its proximity to the business sphere. COBIT strives to create business value. There is nothing worse than doing things for the sake of it and forgetting who you work for (the business). People who forget these things are not value enablers but the exact opposite.

Everything in COBIT reminds you who you work for and why you do it, and that is why I like it. **COBIT for Risk** is an addition to COBIT and is meant to be used together with the generic COBIT framework. The following screenshot illustrates one of the essential parts of COBIT and COBIT for Risk—namely, the seven enablers:

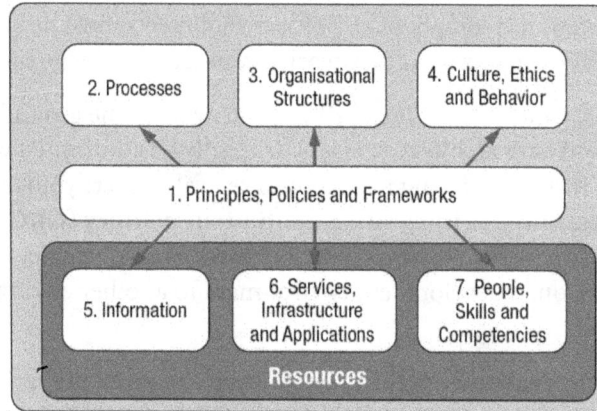

Figure 2.6 – COBIT's seven enablers

These seven enablers are organizational pillars that you should leverage whenever you need to find a solution for ruling things efficiently. Note that you are free to add your own enablers.

Let's go through a short description of each of the seven enablers, as follows:

- **Principles, Policies, and Frameworks**: This could be summarized as *what is clearly thought is clearly expressed*. You should identify your core principles and policies that are in line with the risk appetite of your organization. These will be later shared and reused among all involved parties.

- **Processes**: These represent the actual means to execute policies and transform principles into tangible outcomes—for example, the *MEA03.01* process (where **MEA** stands for **Monitor, Evaluate, and Assess**) helps you tackle changing compliance requirements. COBIT comes with many processes, giving you a strong structural approach.

- **Organizational Structures**: These are key enablers to putting an organization's business and IT goals and risk management practices in motion—for example, hiring a **data privacy officer (DPO)** is a direct enabler to manage **General Data Protection Regulation (GDPR)**-related concerns. The DPO could rely on a few built-in COBIT processes or add their own to get GDPR concerns under control.

- **Culture, Ethics, and Behavior**: The DNA of a company is often embodied within its employees and collaborators, which is good and bad at the same time. It is good because people who share the DNA of the company should strive toward the company's goals. It is bad because if the DNA changes, the same people might become an impediment to driving that change further—for example, switching from a risk-averse culture to a risk-friendly one or the other way around is not easy. People might just keep working the way they have worked for the past decade, which would slow down your change ambitions. Any big change requires a proper change management program next to it. A company should create risk-awareness campaigns to distill a certain mindset within their troops because they are the first line of defense. As a software architect, it is important to understand the company culture to optimize your interactions with peers and managers.

- **Information**: This enabler leverages existing information—for example, you may create a risk register and keep it up to date. This should help you keep good control and oversight of overall risks.

- **Services, Infrastructure, and Applications**: This enabler will simply support the information enabler.

- **People, Skills, and Competencies**: You cannot achieve anything without the right people and the right competencies. When you decide to use a risk management framework, make sure people get to learn it.

The power of COBIT is that you can apply the preceding seven enablers to any topic, from global governance to a scoped risk management function.

In the next sections, I will only highlight the COBIT potential, scoped to the risk function, but I strongly advise you to have a deeper look at it, especially if you are involved in any governance exercise.

Understanding risks

Back on topic, what is a risk? A risk is the probability of an adverse event to occur and have a negative impact on an asset or a company. A risk can, most of the time, be avoided or mitigated. Residual risk is the remaining level of risk once all mitigations are in place. The purpose of a risk assessment is to ultimately evaluate what residual risk remains and let the business make an informed decision about it. A well-conducted security practice aims to inform the business about risks, not imposing anything.

To take an extreme example, a business might consciously decide to violate a regulation if the cost to comply with it is higher than the fine (and other impacts) the company would get when not complying.

Risk management is about identifying, quantifying, and managing risks. It is not restricted to security risks. COBIT 5 comes with a few tools to manage risks, outlined as follows:

- Risk management processes, which help you tackle risk management in a structured way

- Risk taxonomy, which is a tool to evaluate possible risk impacts and their frequency

- Risk scenarios, which are used to analyze risks and propose possible risk responses

Risk-scenario example

A good example is always better than lengthy explanations. COBIT comes with 20 risk categories. Let me use two very different risk-scenario examples, one technical and one non-technical, to show that COBIT can be used to manage any type of risk. Let's start with the non-technical example. The following screenshot depicts a risk scenario about possible sourcing issues regarding a program (group of projects):

Risk Scenario Title	Resource Unavailability
Risk Scenario Category	Portfolio establishment and maintenance
Risk Scenario Description	
Resource allocation is often subject to change due to a switch of priorities across projects of program XXX.	
Risk Scenario Components	
Threat Type: Failure in resource allocation	
Actor: Internal resource allocation is managed by the program manager in cooperation with the IT leadership team.	
Event: Ineffective execution of *APO05 - Manage Portfolio process*	
Asset/Resource (Cause): Process APO05 is not well applied.	
Duration: Extended because the new system might live with important delays	
Occurrence: Timing is noncritical	
Detection: Slow as the key resources will progressively be made unavailable with new coming priorities	
Lag: Delayed as the effect of staff unavailability will be visible at a later stage.	
Possible Risk Response	
Avoid: Pre-allocate resources and do not switch later.	
Accept: N/A	
Share/Transfer: N/A	
Mitigate: Re-prioritize the program and ensure knowledge transfer in case of resource switch.	

Figure 2.7 – Non-technical risk scenario

The risk scenario depicted in *Figure 2.7* is part of the **Portfolio establishment and maintenance** category, and the scenario depicts an issue with resource allocation. As you can see, this risk is not security-related. Conversely, the following screenshot shows the same risk-scenario structure applied to a technical topic:

Risk Scenario Title	Symmetric key compromised
Risk Scenario Category	Infrastructure theft or destruction
Risk Scenario Description	
Symmetric keys could be compromised if stored improperly or used maliciously, leaving the door open to unattended access to Service Bus.	
Risk Scenario Components	
Threat Type: Malicious	
Actor: Internal, malicious insiders (most likely). External, Attackers (less likely).	
Event: Theft	
Asset/Resource (Cause): Process BAI10 is not well applied.	
Duration: Extended	
Occurrence: Timing is noncritical	
Detection: Most probably slow	
Lag: Immediate as the attacker can immediately authenticate against service bus	
Possible Risk Response	
Avoid: Use Azure AD to authenticate and prevent access to *RootManageSharedAccessKey*	
Accept: N/A	
Share/Transfer: N/A	
Mitigate	
• Store keys in Key vault • Renew keys frequently • Monitor the *get secret* and *list key* operations	

Figure 2.8 – Technical risk scenario

The risk scenario depicted in *Figure 2.8* describes the possible malicious usage of a service bus symmetric key, which is used to authenticate against the bus. Parameters such as duration, detection, and lag help qualify the risk and identify the time between its occurrence and its consequences. These two very different examples show how you can use COBIT to your advantage for risk management. Both risks refer to COBIT's predefined categories and processes. You can extend COBIT within your categories and processes. Combining risk scenarios with the other COBIT tools, such as leveraging COBIT enablers, represents a powerful toolbox.

COBIT wrap-up

As a software architect, you might be confronted with highly sensitive applications and systems. You could use risk scenarios since they are straightforward and nice to formally describe and respond to risk. We have not seen the other COBIT tools, but feel free to explore these further on our own. Now that I have briefly introduced COBIT, let's look at NIST.

NIST

NIST is a well-known **United States (US)** framework (`https://www.nist.gov/`) that you can leverage for any technology-related topic. It ships with a cybersecurity-specific framework—namely, **NIST Cyber Security Framework (NIST CSF)** (`https://www.nist.gov/cyberframework`). NIST CSF covers all the duties I introduced earlier when describing the CISSP certification. As with COBIT for Risk, NIST CSF also handles risk management, but both frameworks can be used together. We will not look at NIST any further, but I wanted to make sure you keep it on your radar because the NIST organization publishes articles, reference architectures, and so on that might be useful for a software architect. I personally used such publications for cloud-related topics in the context of a cloud strategy exercise. To complete our framework journey, let's look at the indestructible ITIL framework.

ITIL in a nutshell

ITIL is a **United Kingdom (UK)**-born framework. Back in the 1980s, the UK administration realized that its internal IT service management was rather chaotic. They decided to clean the house and build a framework that would assemble best practices from all over the place when it came to service management. Since then, ITIL, currently at **version 4 (v4)**, is a world-leading IT service management framework. Infrastructure and operations teams around the world use ITIL, consciously or unconsciously.

ITIL primarily focuses on internal customers. For example, when you, as an employee or collaborator, request a corporate laptop to work with or when you need to access a certain system, you resort to ITSM. ITSM encompasses the organizational capabilities that deliver value to a customer (internal, in this case). The infrastructure department and the service-desk function are the back offices of a company. They keep the lights on and make sure everyone has the necessary tools and access to work. This is often thankless work but is necessary for a company to operate.

In 2018, ITIL v4 superseded the long-lived ITIL v3, which many organizations are still using today. ITIL v4 has changed a few concepts and extended the ITIL practices beyond ITSM capabilities. However, typical capabilities (request management, problem management, incident management, and so on) that all organizations need are still in scope.

As a software architect, you are more likely to contribute to application development to respond to **functional requirements (FRs)**. You should also be partially involved in NFRs (the subject of our next chapter), which include typical topics such as scalability, **high availability (HA)**, and so on. Many NFRs are not only code-related but also infrastructure-related, so you will *need* to collaborate with infrastructure architects. You might have experienced some mindset clashes and siloes between development and infrastructure teams. This is partly due to the way these teams are organized. Most development teams have adopted agile for a long time already, while infrastructure teams remain organized around ITIL. In ITIL, nothing is improvised; everything is thought through upfront and follows a clear life cycle and clear processes.

As a software architect, you should grasp the essential parts of ITIL to better interact with infrastructure architects. ITIL makes you understand that infrastructure teams deliver transversal services (mail, document management, network, corporate devices, and so on) that are scoped to the level of the organization, while most applications have a narrower scope. The scale of magnitude of these services partially explains why they rely on robust service management principles because they simply cannot afford an outage. In essence, ITIL is about service design and service operations. ITIL is about *measuring* and *managing* your IT services.

ITIL remains a fundamental pillar of the IT landscape, but DevOps and DevSecOps try to claim some space to make collaboration and toolchains more efficient across teams. ITIL promotes customer satisfaction, but real-world implementations are (too) often very IT-centric, while agile is more business-centric. The goal of DevOps and DevSecOps is to bridge these views, both of which are necessary.

Summary

In this chapter, we browsed through some of the most widespread frameworks in the IT industry. You learned that TOGAF and ArchiMate are the languages of enterprise architects. As a software architect, you might be brought in to draw architectural patterns and some EA building blocks. We then reviewed NIST and COBIT for Risk, which come in handy to drive a security practice. We finally explored ITIL, the de facto ITSM framework used by most organizations. By understanding the essential parts of these frameworks, you should be able to optimize your interactions with other stakeholders. These extra skills might become a differentiation factor between you and an average software architect. Of course, I encourage you to explore some of the frameworks further, as these were only introduced in this chapter.

In the next chapter, I will introduce a software architecture methodology that I have been using at different places and that you will undoubtedly encounter sooner or later.

3
Understanding ATAM and the Software Quality Attributes

This chapter will provide you with an overview of the **Architecture Tradeoff Analysis Method (ATAM)**, which is a widely adopted architectural analysis method used in organizations. As explained in the introductory chapter, software architecture practice may vary from one corporation to another or from one industry to another, but you should encounter ATAM on your way sooner or later. In this chapter, we want to make you understand ATAM's essentials.

We will more specifically cover the following topics in this chapter:

- Introducing ATAM
- Understanding sensitivity points, trade-off points, risks, and non-risks
- Exploring quality attributes

- Getting started with quality-attribute scenarios
- Practical use case
- ATAM and agile at scale

Introducing ATAM

Design choices are all about trade-offs. In many enterprises and for many projects, we usually aim to design and develop top-notch software but may end up with unexpected outcomes. These deviations from initial expectations could be due to shortcuts we were forced to take, budget restrictions, a permanent scope change, a lack of proper analysis, a lack of a well-thought-through architecture, and so on. All these reasons may lead to some design choices that in turn lead to trade-offs. Without a formal way of identifying these trade-offs, organizations lack the ability to make informed decisions or even to conduct **root cause analysis** (**RCA**) when problems occur in production.

An example of a shortcut could be that there is no budget left to fine-tune some security aspects, but in omitting this, you will potentially increase your exposure to malicious users. For an asset that deals with public data, this would be less risky than for an asset dealing with **personally identifiable information** (**PII**) data. Similarly, if the number-one requirement of a system is to be portable, you would likely envision a containerization platform instead of writing a plain .NET application leveraging Windows-specific features.

As you can see, it's all about exercising good judgment over a specific context, and in essence, that is what ATAM is all about. **ATAM** helps software architects assess both some generic and specific concerns by assessing design choices against software quality attributes to meet some quality goals.

This may sound obvious, but quality, often defined as a combination of *fit for purpose* and *fit for use*, is sometimes an absolute requirement for software to function correctly or to adhere to high industry standards. While a slow report may irritate an end user, in a true real-time system, not being able to handle an event within 3 milliseconds would just not be acceptable.

Let's come back to the fit-for-use and fit-for-purpose notions. **Fit for purpose** means that what was developed corresponds to the functional requirements, while **fit for use** means that what was developed works reliably. A typical example to illustrate these notions is given here: What do you expect from a washing machine? You expect it to clean your clothes *and* to function reliably. Indeed, a machine that would clean optimally but would encounter breakdowns every 2 weeks would be disappointing. That is why quality is the combination of these two notions.

The beauty of ATAM is that it forces you to think about fit for use and fit for purpose, which influences the resulting quality of a product.

ATAM will help you define your quality goals and make design decisions accordingly. It makes it possible to detect whether a quality attribute of interest is affected by one or more design decisions, enabling you to set priorities and focus on the most important factors that will be key to the success of the product you are building. ATAM should ideally be performed prior to the actual implementation, but you can also use it in *brownfield scenarios* to assess the adequacy of an as-is architecture. ATAM should be used as an inspirational source and in a pragmatic way. You do not need to apply it by the book.

Let's now explain the heart of ATAM.

Understanding sensitivity points, trade-off points, risks, and non-risks

As stated in the previous section, making a design choice often has a positive or negative impact on one or more quality attributes. The two most important concepts to grasp are sensitivity points and trade-off points, and, to a lesser extent, risks and non-risks. Let's have a quick definition of these before going into more elaborate scenarios, as follows:

- **Sensitivity points** represent architectural choices that can importantly impact a single quality attribute—for example, maintenance activities could impact the availability of a quality attribute.

- **Trade-off points** are architectural decisions that may impact at least two sensitivity points, hence the trade-off, meaning that you might be obliged to sacrifice one quality attribute in favor of another one. For example, doing client-side encryption will impact security in a positive way but performance in a negative way. You may opt for better security but inferior performance.

- **Risks** are the consequences of some architectural decisions. You and the business might accept a certain level of residual risk—for example, you may decide to host your asset in a single data center. The residual risk is that you may experience an entire outage, should the data center suffer from a major disaster.

- **Non-risks** are opportunities brought by *good* architectural decisions—for example, using a certain technology may bring you additional security features out of the box.

Sensitivity points and **trade-off points** tell software architects where to focus their attention. They help identify what is important to achieve quality goals. Remember that these goals should be derived from business goals. Here are a few examples of sensitivity points:

- **Scalability**: The level of parallelism *might be sensitive* to the number of queues and/or partitions defined in the message broker.

- **Performance**: Latency *might be sensitive* to the application protocol we choose (for example, **HyperText Transfer Protocol 1 (HTTP/1)** or HTTP/2).

- **Security**: The risk of data leakage *might be sensitive* to the level of encryption applied to data at rest.

  It is good to use the *might be sensitive* or *is sensitive* phrase construct when expressing sensitivity points, to emphasize the relationship between the consequence and the cause. Once defined, sensitivity points act as a baseline of factors that could impact the resulting quality. So, for instance, when taking the second item in the preceding list, there is a world of difference when working with HTTP/1 or HTTP/2 in terms of latency (better performance). However, although latency is primarily concerned by the protocol in use, other items, such as load balancing, could also be affected. Let's add an extra sensitivity point to illustrate our purpose, as follows:

- **Availability**: Backends *might be sensitive* to the application protocol we choose (for example, HTTP/1 or HTTP/2).

With the preceding sensitivity point added, we have two different quality attributes (performance and availability) that could be significantly impacted by the choice of protocol we make. That is a *trade-off point* because opting for HTTP/2 has a positive impact on performance (improved latency) but a potentially negative one on availability. The rationale behind this is that most load balancers are still layer-4 (**Transmission Control Protocol/Internet Protocol (TCP/IP)**) and do not understand layer-7 (in our case, HTTP/1 or HTTP/2), meaning that they will fail to properly load-balance HTTP/2 (used under the hood by the popular **Google remote procedure call (gRPC)**) because existing clients will keep sending their requests to the same backend, irrespective of whether that backend is healthy or on the verge of becoming unhealthy.

So, to make a proper design decision in this case, we would need to also consider our load balancers and make sure they are compatible with such a decision. If they are not, we might reconsider the protocol we want to use. If we keep using HTTP/2 despite our old-school load balancers, we end up with a **risk**. ATAM defines risks as potentially problematic design decisions or decisions that have not been made yet. On the contrary, a **non-risk** is based on some assumptions.

For example, on the same topic, an example of a non-risk would be this: Assuming a concurrent user base of 10 users per minute, our backend systems should still be able to deal with the load, despite our layer-4 load balancers' lack of proper understanding of layer-7 protocols. This non-risk is, therefore, valid while the assumption remains valid. Should the expected user base grow significantly, this non-risk will not be valid anymore or must be revalidated.

Another possible non-risk could be implied by some architectural styles; for instance, the system will always talk to the most responsive backend service because our service mesh technologies are layer-4 and layer-7 aware, under normal or high load.

If you have understood this section, then you have understood the essence of ATAM. It is a method that enables you to identify the key quality goals and success factors of well-crafted software architecture that is fit for purpose and fit for use. Let's now explore the quality attributes.

Exploring quality attributes

When building an application or a system, you are always confronted with both **functional requirements (FRs)** and **non-functional requirements (NFRs)**. FRs are emitted by business users and business analysts, and they describe the features that must be developed or made available. NFRs are most of the time expected, but not directly expressed by the business. Indeed, every user wants to have a performant, reliable, and always available application. Similarly, every user assumes that the application/system is safe and respects their privacy, but all of this comes at a price, and the extent to which you want to respond to these NFRs will impact the cost and time dimensions. The systems you integrate with, the hosting platform you choose to host the asset, and many other factors could make it hard to achieve the NFRs.

The purpose of quality attributes is to express the NFRs and define expectations accordingly. The list of software quality attributes is broad; on its `https://en.wikipedia.org/wiki/List_of_system_quality_attributes` web page, Wikipedia regroups about 80 attributes. It is, therefore, crucial to refine them and pick the ones that make the most sense to you and your specific business scenario or industry.

Indeed, in some industries, you might have some common ground across assets from which you can extract a quality attribute baseline. In any case, whether you have a baseline or not, a good starting point is to request the **service-level agreement (SLA)** expected by the business. In large organizations, such SLAs are mainstream, even for internal customers.

As part of the SLA, there is usually the notion of a **Recovery Time Objective** (**RTO**) and a **Recovery Point Objective** (**RPO**). In a nutshell, these two objectives say a lot about the expected availability and level of resilience. With an RTO of 2 business days, we clearly understand that the business is not too concerned by a system outage. Conversely, an RTO of 30 minutes directly sets a high priority to the availability quality attribute. The same applies to RPO, where the business could afford to lose some data— for example, when data is being replicated from a master system. With low RPO requirements, you already know that you don't especially need to invest in an **always-on Structured Query Language** (**SQL**) **cluster** or in a data service that offers **point-in-time restore** (**PITR**) capabilities. RTO and RPO are very structured questions.

On top of RTO, RPO, and an SLA, you should also try to anticipate business growth, **time to market** (**TTM**), the **total cost of ownership** (**TCO**), and so on to refine and prioritize the quality attributes for your product. Note that cost concerns are not a part of ATAM, but we all know that budget is key in every project.

Over the 80 identified quality attributes on Wikipedia, the following are very commonly seen:

- **Usability**: You can see usability as a way to evaluate whether the application/system is fit for purpose. This is one of the rare quality attributes that focus on functionality.

- **Availability and reliability**: Both attributes are somewhat related. The availability of a system is expressed in the number of nines—for example, the highest SLA for availability is 99.999, meaning that the system is available 99.999% of the time. No one is foolish enough to commit to 100% availability. Most of the time, having 99% or 99.9% availability is already very good. That said, 99.9% availability means that you can afford downtime of about 8 hours 45 minutes per year, but it does not mean that you can be out for 8 hours 45 minutes in a row. That is where the RTO comes into play. Reliability reflects the capacity of a system to operate as defined. You should not observe different behaviors when performing the same action. A lack of availability has an intense impact on reliability.

- **Responsiveness**: This attribute relates to performance. A pragmatic way to evaluate the responsiveness of a system is to measure its latency.

- **Maintainability**: This attribute reflects the capacity of a system to absorb bug fixes as well as change requests.

- **Scalability**: This attribute reflects the capacity of the system to handle extra load. Note that in most modern applications and systems, they are often expected to have at least an on-demand scaling rather than a pre-scale up/out. Scaling out means adding extra instances of a service/component, while scaling up means allocating more memory and **central processing unit** (**CPU**) power to each instance. In serverless architectures, you would even refer to elasticity, which means that the system dynamically scales out/in automatically, so as to pay for the exact resource consumption and without having to even worry about it (except for the cost, of course).

- **Testability and deployability**: Nowadays, many enterprises have adopted, or are on the verge of adopting, agile methodologies (agile at scale). This change in the way projects are conducted has an impact on the pace at which new features are developed and released to production. Because of this, it is usually required to have solid foundations (**continuous integration/continuous delivery**) (**CI/CD**) platforms) that make this possible. The higher the number of releases, the more tests must be performed. Thus, the testability of the system becomes very important too.

- **Portability**: This attribute requires that the system should be able to work in different environments.

- **Security**: This attribute is not part of Wikipedia's list but is definitely important, especially in our hyper-connected world and the ever-expanding cloud footprint encountered in many enterprises. The purpose of the security attribute is to make sure the system is not vulnerable to both internal and external malicious or accidental actions.

ATAM is not prescriptive in terms of which quality attributes you should or should not use. The chosen quality attributes will vary according to your use case and your priorities. Let's now explore quality-attribute scenarios.

Getting started with quality-attribute scenarios

Once you have identified a list of quality attributes applicable to your system, you are ready to define quality-attribute scenarios. The purpose of these scenarios is to connect the dots between the events, their outcome, and the expected answer from the architecture. ATAM describes the following three types of scenarios:

- Use case

- Growth

- Exploratory scenarios

Most of the time, only use case scenarios are considered. They intend to describe possible use cases and the expected response from the architecture. Growth and exploratory scenarios aim to anticipate what could come next. The selection of the appropriate scenario should be done in agreement with the relevant stakeholders, including product owners and project managers. While ATAM provides tools such as **utility trees** to translate business goals into quality attributes, the reality shows that we often end up with an Excel sheet (or Word document) regrouping all scenarios and discussed mostly within an architecture. Nevertheless, we will see an example of a utility tree in our next section.

Practical use case

In this section, we will go through a very simplified use case for the sake of brevity. Here is the business scenario:

> Contoso needs to provide a data upload channel for its customers. Uploaded data files may contain errors and must go through a data-cleansing and data-wrangling phase. If errors cannot be automatically fixed, an error file should be returned to the sender through a callback notification. After this first data-check and transformation phase, the resulting validated data is routed to the relevant systems for further handling. In addition, customers require that data can only be processed in Europe for sovereignty reasons, and they are used to File Transfer Protocol (FTP) systems.

> The upload channel should always be available, while the actual handling of the data might be deferred in case of a system outage. It is expected by Contoso's customers to be notified back within 24 hours. The total volume of data uploaded by customers is about 1 terabyte (TB) per month and, during peak times, up to 250 customers may upload data at the same time. For regulatory reasons, the retention time of original files sent by the customers is 5 years.

From this simplified use case, the business has already provided a few important things, outlined as follows:

- The upload channel should always be available.
- Up to 250 customers may upload data at the same time.
- Data should be ingested and processed in Europe. It should also be stored for 5 years.
- The **end-to-end (E2E)** file treatment time should not exceed 24 hours.

Let's now derive a utility tree out of our findings.

Utility trees

For the sake of brevity, we will skip the typical quality attributes, such as usability, deployability, and maintainability, that would normally be part of the exercise, whether explicitly expressed or not by the business.

Let's focus on what was explicitly expressed, and craft the following utility tree:

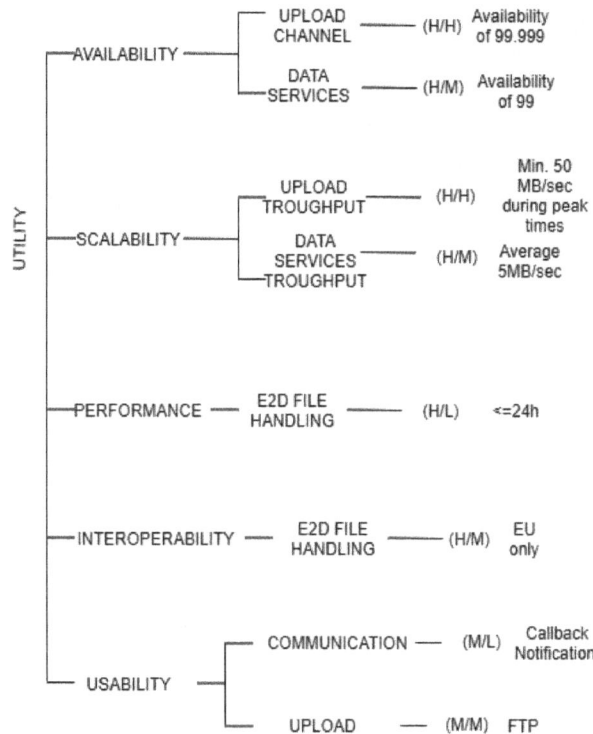

Figure 3.1 – Utility tree

Figure 3.1 represents a simplified view of what we could capture out of our business scenario. Each leaf of the tree represents an **architecture significant requirement** (**ASR**). We decided to focus on availability, scalability, performance, and interoperability. Regarding **availability**, we focused on both the upload channel and the data services. As you can see, the upload channel is marked with a **(H/H)** annotation, where **H** stands for **High**; possible values are Low, Medium, and High. This quantification represents the impact on the business (left side) and on the architecture (right side).

This concretely means that this requirement is highly important for the business and is considered high-risk for the architecture. To put this differently, this will not be our low-hanging fruit and we will have to pay close attention because it will not be easy to achieve.

Conversely, we find the E2E file-handling underperformance and we consider it low-risk. We also retained **scalability** because we already know that the system will be confined to peak times. **Performance** is a part of the mix because we have a time expectation from the customer. Finally, **interoperability** highlights the fact that all the components must be located in Europe. In normal circumstances, the utility tree would be much richer than what is illustrated in *Figure 3.1*.

Nevertheless, following the utility-tree generation, we identified a few quality attributes as being the main business drivers for our architecture—namely, availability, scalability, interoperability, usability, and (to a lower extent) performance. What is key to remember is that all stakeholders should share a common understanding of the quality attributes that were chosen. For example, some stakeholders might be tempted to categorize data sovereignty somewhere under security and that's all fine, as long as everybody is on the same page.

In other words, the following conclusions can already be made:

- The system is at least partially working if the upload channel is available.
- The system is partially working if the upload channel is available and the data services are down.
- The system is fully working if both the upload and data services are available.
- The system is performing correctly if the data sender receives an answer within 24 hours.
- The hosting platforms for our components must be located in Europe.
- Most customers traditionally send files through FTP.

The interesting piece in the preceding observations is that we have identified our mission-critical service: *the upload channel*. It is always important to identify such services because we cannot reasonably think that we will be offering 99.999% for every service/component of a solution. Aiming at such availability would undoubtedly cost a fortune, which most projects cannot afford; although cost per se is not considered by ATAM, we all know that in the real world, it is a very important factor. Now we have our main drivers, let's try to identify some scenarios.

Quality-attribute scenarios

In some organizations, you might see that some architectural approaches are already on the table before the scenarios are known. While it is, of course, possible to identify scenarios later, this might result in biased scenarios. It is up to you to decide in which order you want to roll that sequence out, but going for the first run of scenarios is probably a better approach.

A quality-attribute scenario has the following properties:

- A description
- A quality attribute
- An environment that represents the context
- A source of stimulus, which is the trigger for an event (stimulus)
- A stimulus, which is the event itself
- An artifact, which is the concerned component
- A response, which is the expected system behavior
- A response measure, which is the measured behavior

You are likely to see the source of stimulus and the stimulus consolidated, as well as a response and response measures. The environment is also often skipped. The reason for this is sometimes a lack of time. Most quality scenarios are depicted in Excel sheets or Word documents. Sometimes, you may even end up with the following structure:

- A description
- A quality attribute
- Sensitivity points and trade-off points
- An architectural approach
- Risks and non-risks

The preceding structure combines both the questions, sensitivity points, and trade-off points, as well as potential solutions/mitigations in one shot. This could be perceived as some sort of *ATAM express*. Remember that what is important is to identify sensitivity points, trade-off points, and risks.

Let's go through some quality-attribute scenarios for our use case. We will infer them from our utility tree created in the previous section, but for the sake of brevity, we will limit ourselves to six scenarios, as follows:

S1 – Description	Operations on the platform must happen during planned downtime. The impact over the upload channel(s) should be minimal and short.
Attribute	Availability.
Business value/ architectural impact	High/Medium.
Stimulus	Maintenance tasks.
Environment	Underlying upload channel infrastructure.
Artifact	Upload channels.
Response	Operations occur within an agreed timeframe and maintenance takes less than 2 hours.

Table 3.1 – First scenario

S2 – Description	Upload channel(s) outage
Attribute	Availability
Business value/ architectural impact	High/High
Stimulus	Outage
Environment	Channel hosting platform
Artifact	Upload channel
Response	Automatic restore of upload channel within < 10 minutes

Table 3.2 – Second scenario

S3 – Description	Data service(s) outage
Attribute	Availability
Business value/ architectural impact	Medium/Low
Stimulus	Outage
Environment	Data platform
Artifact	Data services
Response	Degraded user experience (UX) until service(s) is/are back up and running, with an RTO of a maximum of 6 hours

Table 3.3 – Third scenario

S4 – Description	Data sent by customers must be handled E2E within Europe.
Attribute	Interoperability.
Business value/ architectural impact	High/Medium.
Stimulus	File transfer and handling.
Environment	At runtime.
Artifact	Customer data file.
Response	Requests are treated in Europe 100% of the time.

Table 3.4 – Fourth scenario

S5 – Description	Customers prefer to send files over FTP.
Attribute	Usability.
Business value/ architectural impact	Medium/Medium.
Stimulus	File transfer.
Environment	At runtime.
Artifact	Customer data file.
Response	FTP Secure (FTPS)/Secure FTP (SFTP) channel(s) is/are exposed through port 21 or 22.

Table 3.5 – Fifth scenario

S6 – Description	During peak times, there could be up to 250 concurrent customers sending data files.
Attribute	Scalability.
Business value/ architectural impact	High/High.
Stimulus	Increase in memory, CPU, and input/output (I/O).
Environment	In normal circumstances and during peak times.
Artifact	Upload channel.
Response	Provisioned channels will be able to ingest 5 megabytes (MB)/second; additional capacity will be made available to accommodate 50 MB/second during peak times.

Table 3.6 – Sixth scenario

Many other scenarios could be crafted, but the preceding six scenarios are among the most relevant. We skipped the performance aspect since there is no real hard constraint on that one, and we left out security.

Identified sensitivity points

Out of the scenarios listed in the previous section, we can infer a few sensitivity points, as follows:

- **SP1**: The availability of the upload channel *is sensitive* to the amount and duration of operational maintenance.
- **SP2**: The availability of the upload channel *is extremely sensitive* to the level of redundancy.
- **SP3**: The scalability and, especially, the maximum ingress throughput *is sensitive* to the number of concurrent upload channels.
- **SP4**: The usability *is sensitive* to the upload method.
- **SP5**: The interoperability *is sensitive* to the hosting platform location.

Now we have our list of scenarios and a few sensitivity points, let's try to respond with some architectural approaches and evaluate their impact. We should see trade-off points emerging after a first analysis of the proposed architectures.

Architectural approaches

Before diving into different approaches, let's look at the following simplified diagram showing the different components and the flow:

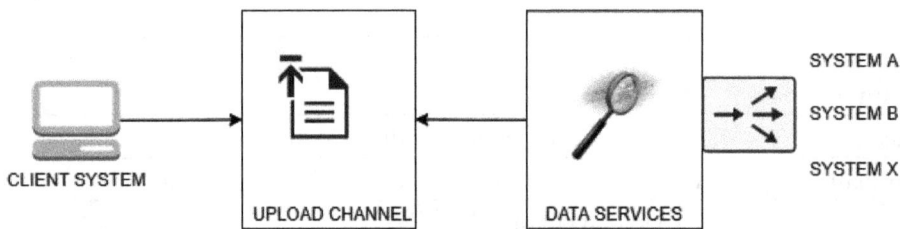

Figure 3.2 – Our simplified component diagram

From left to right, the client system sends data files to the upload channel. Files are picked up by our data services for data cleansing and data wrangling. The resulting data is routed to different systems according to the nature of the input data. Note that we completely left out the integration with the other systems, which, in reality, would certainly have a big impact on the architecture. Let's focus on the data ingestion and data handling aspects.

Approach A – On-premises infrastructure

Contoso already has some on-premises SFTP servers that could be used to ingest the customer data. Contoso is a US-based company and has one data center in Europe. The Contoso team brainstorms and comes back with the following proposal:

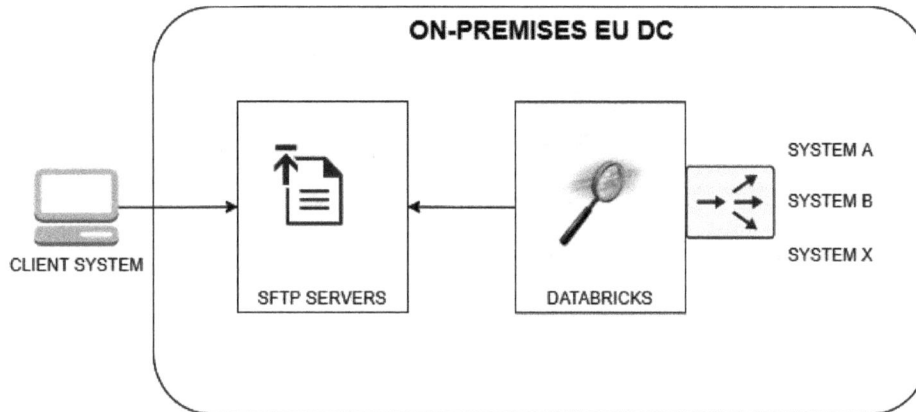

Figure 3.3 – High-level architecture proposal: first option

On top of having SFTP capacity, they also have on-premises Databricks clusters that could handle the data-cleansing and data-wrangling operations. Reviewing the quality-attribute scenarios, they realize a few potential impacts over the previously identified sensitivity points, as follows:

- **SP1**: The maintenance window to patch the SFTP servers is very short, and it will be challenging for Contoso to find a path and operate this potentially large number of servers.

- **SP2**: Because Contoso has only one data center in Europe, they are not able to mitigate a **disaster recovery** (**DR**) should a major data center incident occur. Availability is therefore at risk because there will be no redundancy in another data center.

- **SP3**: They still need to do their math, but they find it challenging to be able to accommodate an ingress of 5 MB/second and 50 MB/second during peak times. On top of this, they realized that 1 TB per month over 5 years (retention) means 60 TB of storage. They have to double-check with the storage team, but it sounds huge at first sight.

A trade-off point (**TP1**) is identified: they could achieve **SP1** and **SP2** if they could also rely on their US data center. Unfortunately, this has a negative impact on **SP5**.

After having checked with their legal department, it appears clear that **SP5** is an absolute requirement, so **TP1** is not allowed. Thus, they came up with approach B.

Approach B – Cloud-based architecture

Because Contoso's capacity with only one data center in Europe is compromised, they started to envision a cloud-based (Azure) architecture and came up with the following proposal:

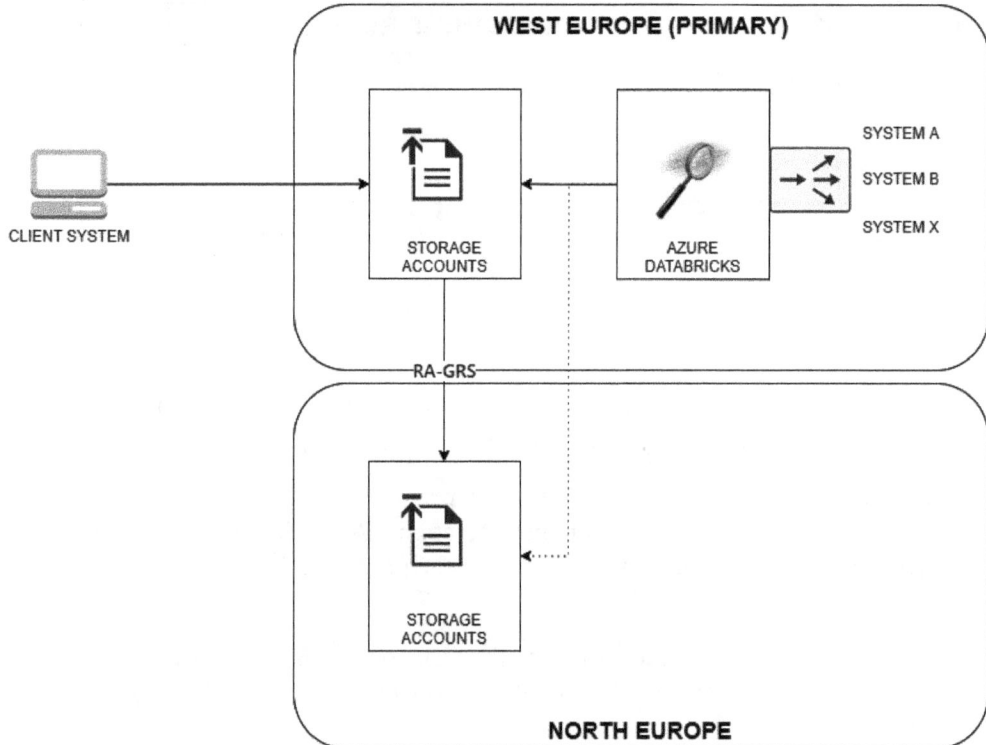

Figure 3.4 – High-level architecture proposal: second option

This time, they leverage two different data centers in Europe. They realize that this proposal is in line with most sensitivity points. While reviewing, they realized the following:

- **SP1**: Using fully managed storage accounts is a **non-risk** because the maintenance is completely transparent and performed by the cloud provider. Moreover, thanks to the activation of **read-access geo-redundant storage (RA-GRS)**, the data is replicated to the secondary region and a failover could be initiated should a regional outage (a rare event) occur. However, RA-GRS-enabled accounts still offer an availability of 99.99%, not 99.999% as expected, so we have a **risk** here.

- **SP2**: A storage account would represent a single upload channel with high support of concurrent uploads. The redundancy is baked into the service.

- **SP3**: Documentation reports that a single storage account may handle a throughput of 10 **gigabytes per second** (**GBps**), meaning about 1.25 GB/second largely above the required throughput.

- **SP5**: They are in line with the obligation to process everything within the **European Union** (**EU**) boundaries.

Although the solution seems more appropriate, they identified a trade-off point (**TP1**): while the upload channel approach has a positive impact over several sensitivity points, it has a negative impact on the usability quality attribute (**SP4**) because customers traditionally use SFTP or FTPS to transfer files.

So, here again, we end up with a trade-off. In reality, you are likely to end up with several trade-off points and residual risks.

Let's now wrap up the use case.

Use case wrap-up

We have to wrap up our use case here, but as you can imagine, similar yet different approaches should go through an analysis, and once all the trade-off points are identified and evaluated, the best approach should be chosen. Possible alternative approaches could have been these:

- A combination of a cloud-based data center and an on-premises data center

- An active-active cloud-based setup with different storage accounts and different Databricks clusters

- Launching a **request for proposal** (**RFP**) to let suppliers submit proposals

We hope that you understood the different mechanisms consisting of analyzing the business requirements, identifying some quality goals, writing business scenarios, and extracting sensitivity points, trade-off points, risks, and non-risks. The objective is to iterate a few times and review multiple options before choosing the target solution. You are likely to end up with residual risks that should be reported to the business.

We have not focused on code-related NFRs such as maintainability, readability, and upgradability because most critical NFRs lie within infrastructure and security and because we will focus only on the code in *Chapter 5, Design Patterns and Clean Architecture.*

Let's now see how to deal with ATAM in an agile-driven organization.

ATAM and agile at scale

As stated earlier, many corporations tend to adopt agile methodologies as part of their digital transformation journey. You may wonder if ATAM is adequate in agile organizations. This question deserves to be asked and probably has no definitive answer.

However, let me try to share my opinion about this. Agile methodologies, in general, aim to deliver incremental business value in a timely and cost-efficient fashion. The important keyword here is **incremental**. Agile embraces the concept of a **minimum viable product (MVP)**, which, in a nutshell, is a *production-grade* application with a minimal number of functionalities that are enough to be considered viable by the business to attract early adopters and launch a product on the market before the competitors. This definition of an MVP clearly targets *functional* features, not really *NFRs*. While this incremental way of working is perfectly possible with features that are yet to be developed, it is hardly applicable to NFRs such as the overall security of a system.

For example, encryption in transit is enabled or is not; data at rest is encrypted or is not. It is hard to find something in the middle that would be considered minimal and acceptable. Security aside, a way out is probably to also apply the MVP concept to the RTO and RPO depicted before. If you manage to lower your expectations, then you can already use ATAM in parallel to craft a more future-proof architecture, to meet the requirements of the full-blown product. That said, agile methodologies often have a very positive impact on code-related quality attributes, such as maintainability, testability, and deployability, because they often rely on strong CI/CD platforms where code scanning and automated testing are first-class citizens.

It is, however, your responsibility as a software architect to tell the business that they will most likely not benefit from a highly resilient architecture at the stage of the MVP. In other words, the target architecture will unlikely be available as of day 1 in a fully agile-driven project/program. You should also aim to increment the maturity level of the architecture.

Summary

After reading this chapter, you should have grasped the essentials of ATAM and its main purpose: discovering sensitivity points, trade-off points, risks, and impactful quality attributes to help you make informed decisions. I strongly encourage you to have a pragmatic (not dogmatic) approach to ATAM because not every asset requires the same level of attention. However, in mission-critical projects, ATAM proves to be efficient and can be credited with raising the right questions at the right time. Flaws resulting from incorrect architectural decisions are very hard to fix and adjust at a later stage. Note that I refer to flaws, not to maturity levels. You can work in an incremental way but you should make sure not to end up with true design flaws in your architecture. ATAM should safeguard you against such adverse situations. In our next chapter, we will review the different architectural styles, from decades ago up to today.

Section 3: Software Design Patterns and Architecture Models

In this section, we will revisit monoliths, service-oriented architecture, and microservices. I will shed some light on the benefits and drawbacks of each architectural style. You will learn why there is a natural evolution toward microservices, but also why you should not bury monoliths too quickly. Although design patterns are not new, a book on software architecture cannot do without them, so I will cover what I think are the most important patterns to master as a software architect.

This section comprises the following chapters:

- *Chapter 4, Reviewing the Historical Architectural Styles*
- *Chapter 5, Design Patterns and Clean Architecture*

4
Reviewing the Historical Architecture Styles

In this chapter, we will review some of the existing architecture styles. I decided to focus on only a few, in chronological order, from monoliths to microservices, but the list of architecture styles does not stop there. There are many more styles and patterns, but the ones I chose represent both legacy and modern systems, which you will definitely encounter in your software architect career.

Here are the topics we are going to focus on:

- Introducing architecture styles
- Starting with monoliths
- Continuing with **service-oriented architecture (SOA)**
- Finishing with microservices

By the end of this chapter, you should be able to understand the benefits and drawbacks of monoliths, SOA, and microservices. You should be familiar enough to recognize the style of the architecture you are confronted with and apply the skills you've gained in your own context.

Introducing architecture styles

Architecture styles are high-level design choices that influence the way applications are designed, built, and hosted. Making such a choice forces you to obey the standard practices that ship with the style in question. Some architecture styles act at a higher level than others.

For example, the **three-tier architecture** is based on three different layers – presentation, business, and data – all of which are *physically* separated. In a three-tier architecture, the presentation layer cannot talk directly to the data layer. Network policies should be enforced to prevent such occurrences. As you can see, this type of architecture has an impact, not only on the hosting piece but also on the way you organize the development of the different components. Conversely, the **Model View Controller** (**MVC**) pattern is also based on three layers, but all the layers can be deployed to a single server. Here, the physical split is not required. However, going for MVC will force you to split your code accordingly, and it will probably push you to use one of the MVC frameworks.

In the same vein, the **Event-Driven Architecture** (**EDA**) and **Publish/Subscribe** models will make you adhere to concepts that are specific to message-driven architectures, such as the **queue-based load leveling pattern**, the **claim check pattern**, and the **competing consumer pattern**, to name a few.

To connect the dots with ATAM, which we discussed in the previous chapter, some architecture styles have a direct influence on quality attributes. They might come with inherent risks and non-risks, which, in turn, may impact the quality attributes positively or negatively. I will highlight this in the following sections, using words in *italics*.

To know whether a given architecture style makes sense in your own context, you must understand the benefits and drawbacks of each. You must also evaluate the capabilities of your organization (or customer) to adopt a certain style, as well as to see how disruptive this style is toward the existing landscape.

The reason why I decided to focus on monoliths, SOA, and microservices is because these styles are still alive today and you are very likely to stumble upon one of these in your day-to-day software architecture practices. Another reason is that they are tightly linked and reflect some sort of chronological evolution of the IT industry. We started, probably unconsciously, with monoliths, then tried to fix monolith-related issues with SOA, to ultimately fix SOA's issues with microservices. What is also remarkable is their related scope. With monoliths, the application is king because you tend to ignore everything that is around it. With SOA, the enterprise is king, and the application must comply with what has been defined at the enterprise level. With microservices, we come back to the application again but refine it as domains and sub-domains.

As you will see in the upcoming sections, each architectural style has a series of advantages and disadvantages. Each style will probably remain in the IT landscape for quite a while, even the so-called monoliths, as we will see in the next section.

Starting with monoliths

I guess that you must already be familiar with monoliths, as it seems they have become the pure evil ones. However, at the risk of shocking you, monoliths will probably be around forever and have some interesting benefits. Before we look at their benefits and drawbacks, let's see what a monolith looks like:

Figure 4.1 – Literal meaning of a monolith

The preceding image shows what a monolith truly is. It is some sort of single-block-rock, from which you cannot extract a single piece, at least not with your hand. It is something that cannot be manipulated easily, something that will have a certain resistance to changes. However, while it seems hard to manipulate the monolith shown in the preceding image, it looks much easier to do with the ones shown in the following image:

Figure 4.2 – Small monoliths

Yes, you got it: *with monoliths, size does matter*! The problem with monoliths is when they grow over time. When you start with a tiny application, it remains possible to adjust it in a timely fashion, without taking on much risk. As time passes by, this little monolith tends to grow and becomes the monolith shown in *Figure 4.1*, which is hardly manipulable. So, how did you get there? Probably because you did not choose an appropriate architecture style, or maybe because you did not follow some basic development principles, or maybe because you did not consider the quality attributes we discussed in the previous chapter.

If you do not enforce layers in your software, or if your layers are tightly coupled with each other, you will inevitably end up with a monolith. That being said, not every application grows over time, and this is why going consciously for a monolith could still be acceptable in some situations. For example, if you have a one-shot job-like program, you may decide to develop it in a monolithic way. This is because you know it will not evolve over time and because it might be the most straightforward way for you to develop it.

At a micro level, from a single application perspective, a monolith somehow promotes agility because it tends to ignore whatever exists in the enterprise landscape. A team developing a monolith does not care at all about what is around; they just do what they need. They do not hesitate to make point-to-point connections with other applications or databases if needed, to achieve project-specific goals. This makes the team gain speed and autonomy, with the risk of becoming the big monolith shown in *Figure 5.1*, should the application grow over time.

At a macro level, multiplying smaller monoliths may result in an enterprise-level monolith, which is way more problematic. If the whole enterprise landscape has become a plate of spaghetti, no one will ever dare to change any of the micro applications anymore, because of an increased risk of cascading side effects. So, as you have probably guessed already, monoliths should be avoided as much as possible. Nevertheless, let's see what benefits they offer.

Benefits of monoliths

It might be hard to believe but monoliths also have benefits. Here are a few of them:

- Developing a monolith is easy. You do not need to learn any software architecture methodology. Their simplicity is probably why we have so many of them.

- They are simple to deploy. Since they are, most of the time, packaged as a single package, redeploying the application is just about deploying a single package.

- They are simple to monitor. Again, monolithic applications are usually deployed on a single machine with a database server next to it. In some cases, you might have multiple frontend/application servers, but there are not too many places to look when you're troubleshooting a problem.

The keyword is *simplicity*, at least as long as the application remains lightweight. Now, let's look at some of the drawbacks.

Challenges of monoliths

Admittedly, monoliths have more drawbacks than advantages:

- Adding new features and fixing bugs becomes complicated. As stated in the previous section, most of the monoliths are packaged as a single package. While this can be considered an easy way to deploy the application, every single feature or bug fix will cause you to redeploy the entire application, which will cause a temporary (hopefully) outage and require many regression tests. This harms *testability*, and ultimately the *deployability* quality attribute. Over time, the team will lack agility, which impacts *upgradability*.

- The code is often tightly coupled. Most of the big monoliths have intermingled classes and subcomponents, which violate the single responsibility principle. This harms the *maintainability* and *readability* quality attributes.

- Lack of granular *scalability*. Because the monolith is a single-block-rock, it must be scaled up/out as one block, although not every part of the application will be used with the same intensity. This causes a waste of infrastructure resources over time.

- *Availability* is at risk. Because the application is atomic, any problem in a specific module could cause an outage of the entire application.

- Higher risk of technical debt. Because changes become problematic over time, teams are hesitant to upgrade to the latest frameworks and technologies. This digs the technical debt deeper and exposes the application to be out-of-support and to security vulnerabilities. The consequence of using monoliths is that they typically have a negative impact on *supportability* and *security*.

The preceding list is not exhaustive, but it gives you a good idea of the risks you take when writing monoliths. However, as we mentioned previously, they could still be a perfect fit for programs/applications that are not intended to grow. Now, let's see how SOA tries to prevent enterprise-level monoliths.

Continuing with service-oriented architecture (SOA)

SOA promotes reusability across the entire enterprise landscape by exposing business capabilities in the form of services. SOA emerged in the last decade of the previous century, with the aim of decoupling applications. Before SOA, it was very common to have client applications directly connect to each other, or to a shared database with read/write permissions. This led to big issues and to the formation of an enterprise-level monolith, as described in the previous section, to the extent that changing anything could pose problems to many applications, leading to a lack of agility and an ever-increasing amount of time required to make small changes.

The following diagram shows a typical SOA implementation:

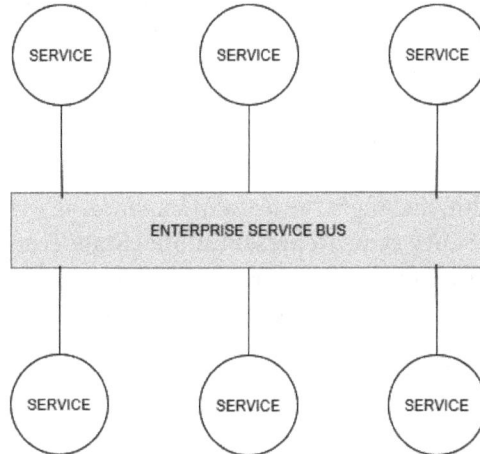

Figure 4.3 – Typical SOA implementation

The component in the middle is the **Enterprise Service Bus** (**ESB**), which plays a central role in SOA. The ESB interconnects different services. Although we can craft the ESB component ourselves, many large enterprises rely on proprietary software such as **webMethods** and **Microsoft BizTalk**. The ESB encompasses duties such as routing and data transformation and holds some business logic. The following diagram is a zoomed-in depiction of the ESB:

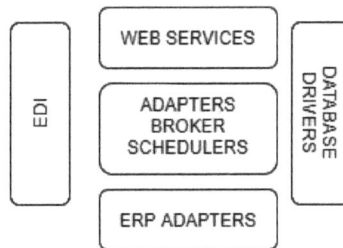

Figure 4.4 – Typical ESB duties

As you can see, an ESB deals with more than just web services. It is used to interconnect every kind of component. The ESB is scaled for the enterprise, which is both a strength and a weakness. Thanks to the ESB, you avoid a chaotic IT landscape with many point-to-point connections, and you ensure proper governance when managing business assets. On the other hand, because it is defined at the enterprise level, it lacks some agility. In some organizations, the ESB itself has become a bottleneck.

Something that eventually played against SOA is **Simple Object Access Protocol** (**SOAP**) for its initial version, then simply SOAP, which was very dominant at the rise of SOA. SOAP was used to exchange data in a structured way. It was supported by **Extended Markup Language** (**XML**), and a web service contract was defined in the form of the **Web Services Description Language** (**WSDL**). While SOAP was initially designed for stateless services, it turned out to be used most of the time for stateful communications when exchanging data. SOAP was also a rather heavy protocol. Its metadata envelopes were consuming a lot of bandwidth, making it the *de facto* standard, as well as inappropriate in some situations. Soon after SOAP came **Representational State Transfer** (**REST**), which went back to the roots of HTTP, to alleviate most of SOAP's issues. However, REST was not designed for SOA, although it is possible to combine RESTful services with legacy SOA.

REST became very popular, while, at the same time, web-based applications and lightweight web APIs became mainstream and even dominant. This rising interest in REST sounded the death knell for SOA. However, SOA is not totally dead yet because companies have invested so much into it, which means it will remain in the air for a while.

Let's take a look at the advantages and disadvantages of SOA.

Benefits of SOA

SOA comes with the following benefits:

- Helps decouple applications and services, which allows for greater interoperability.

- Better governance around IT and business assets. Somehow, SOA aims to provide a single source of truth.

- Promotes *reusability* across the enterprise.

- Helps identify enterprise-grade business assets.

- Technology agnostic, although the systematic use of a proprietary ESB still leads to some vendor locking.

- Improved *scalability* since services are independent and can be scaled according to their own needs.

The preceding list of benefits is not exhaustive, but it is essentially what you gain by implementing SOA. Now, let's review some of the challenges of SOA.

Challenges of SOA

The biggest challenges of SOA are as follows:

- Lack of agility. Because SOA tends to govern the enterprise landscape, and because the ESB is used as its cornerstone, every team and project must integrate/comply with the ESB.

- Increased complexity.

- Increased costs, especially to set up and configure the ESB.

- *Availability* could be at risk. The ESB is a single point of failure. Any outage of the ESB cascades down to all the connected services and applications.

- SOA did not reinvent itself.

Note that these disadvantages are not especially inherent to SOA itself but to the way it is often implemented. Most large enterprises have adopted a top-down SOA approach. Most large enterprises have bought proprietary software from big vendors at a very high cost. Because of this, SOA often leads to a *one size fits all* approach, which is never a good idea, and kills agility. Another factor that plays against SOA is the rise of cloud platforms. SOA is not incompatible with the cloud, but cloud and cloud-native platforms form a paradigm shift, which strongly promotes agility, and that is not completely SOA-friendly. Now, let's explore microservices.

Microservices

Microservices have become popular over the past few years, but it is still not so easy to find a common definition of what they are. In my opinion, microservices can be seen as SOA on steroids, scoped to a single application. Microservice architectures are entirely based on services, but the biggest difference compared to SOA is their level of granularity, their level of decomposition, and their scope. While SOA maximizes reusability across the enterprise landscape, microservices focus on bounded contexts, which may vary from one application to another.

The following is a high-level diagram of what a microservice architecture looks like:

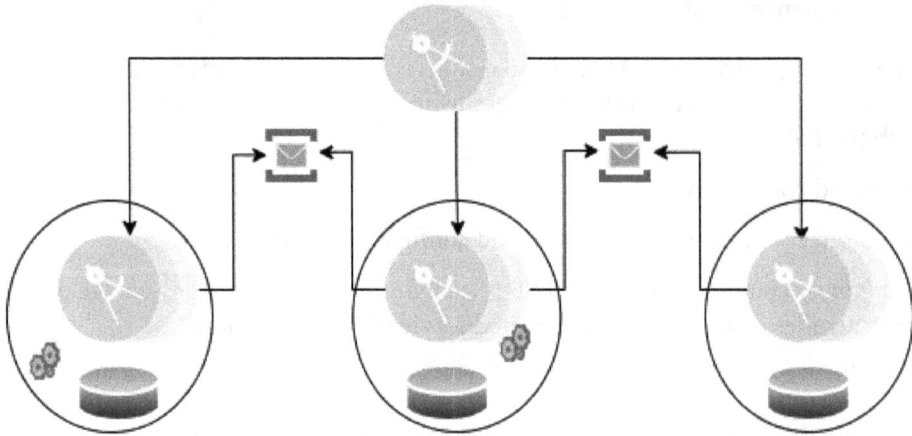

Figure 4.5 – Microservice architecture

Each outer circle in the preceding diagram represents an independent bounded context formed by a microservice. Within a microservice, you may have one or more components and a dedicated data store. Communication across microservices is done asynchronously through the publish/subscribe mechanism. Each microservice potentially exposes a public API to be consumed by clients or other services. Synchronous communication across microservices is discouraged because it may lead to chatty systems, which are less scalable and performant. Synchronous communication should only be used from client apps such as a user interface or a mobile app, or from a **backend for frontend** (**BFF**) because they are user-facing components that are based on a request/response model. Finally, **gRPC** should be preferred for synchronous communication because it is more performant than traditional REST over HTTP/1 communication. Note that REST over HTTP/2 is as fast as gRPC but still less performant in terms of data serialization, which is based on **protocol buffers** (https://developers.google.com/protocol-buffers).

Contrary to the preconceived idea, the size of individual services does not matter. They tend to be small because they have a single responsibility, but a nano service could perfectly violate the microservice principle, should it hold too many responsibilities or deal with matters that belong to another domain.

The term bounded context comes from **domain-driven design (DDD)** and is an explicit boundary of a microservice. This boundary is part of the application where every entity and model is commonly understood by every team member and by the business. This notion of a bounded context is, by the way, the reason why microservices are scoped to a single application: because the same entity might be perceived differently according to a specific application domain. For example, the notion of a customer varies according to whether you are in a **business to consumer (B2C)**, **business to business (B2B)**, or **business to enterprise (B2E)** context, and the same company might be active in all these contexts at the same time. In my opinion, you can do DDD without microservices, but you cannot do microservices without DDD. Doing microservices without DDD will lead you to design a *distributed monolith*, which is already better than a monolith. Now, let's see what the benefits and drawbacks of microservices are.

Benefits of microservices

Because a microservice should be seen as an independent unit of work, the following benefits emerge:

- Greater *deployability*. One of the top benefits of using microservices is that you can deploy them independently, which makes it easy to release them while minimizing potential impacts on other services.

- Greater autonomy. Because microservices have their own databases and their bounded contexts, they benefit from certain isolation. Services that offer an API must remain backward compatible.

- Greater *usability*. When microservices are used together with DDD, the business and the developers work hand in hand to define the domain and sub-domains of the application. The resulting application is often *fit for purpose*.

- Greater *resilience*. Because business activities are scattered across services, an outage of a given service should not hinder other services, or to a lesser extent.

- Greater *scalability*. Thanks to the granularity that's obtained in microservice architectures, we can easily scale out/in services independently from each other, which is cost-effective. You can consider microservice architectures as a *non-risk* from a scalability perspective.

- Greater technology landscape. Microservices are polyglot, meaning that every service may decide to use its own set of technologies and data stores. However, I would advise you to keep this to a manageable level, from an enterprise viewpoint. Microservices based on containers positively impact the *portability* quality attribute.

- Work can be distributed in a more granular way. Each microservice can have its own team.

While the list of benefits is great, microservices also come with numerous challenges.

Challenges of microservices

On the flip side, microservices come with some serious challenges:

- Identifying the appropriate bounded context is probably the most challenging part of microservices. Having a context that's too small may result in anemic services with no business value. Having a bounded context that's too broad may result in encompassing too much business logic and too many concepts into a single context, thereby increasing the risk of it becoming a monolith.

- Due to the distributed nature of microservices, a single transaction often involves multiple services, leading to orchestrator-based or choreography-based sagas (a sequence of actions), which increases complexity.

- Because microservices hold their own databases, you might have to react to data events occurring in other bounded contexts in order to replicate what you are interested in locally. This leads to **eventual consistency**, which means that the data will only become consistent after a certain period. Eventually, consistency systems are often built to optimize read operations at the cost of accuracy. This is not the time to replicate some data changes from one bounded context to another since this leads to an eventually consistent system. This may not be suitable in every situation. For instance, if you want to know the balance of a bank account, you want to make sure you get it right the first time. Conversely, the number of likes you received for your last blog post does not matter too much, so you do not need to have the absolute truth when you refresh your blog post page.

- Potential *performance* impacts due to distribution tax. While in the monolith world, a single call is enough to do the entire job, in a microservices world, you often end up with multiple calls, potentially leading to network exhaustion, or at least deteriorated performance. Patterns such as the circuit breaker help prevent such network saturation. The best is option, of course, to rely on asynchronous patterns as much as possible.

- Loss of complete oversight. Due to the dilution of responsibilities and ownership, it becomes harder to grasp which service does what as an application grows.

- Harder to monitor. Again, because services may use different technologies and data stores, it is often more challenging to monitor them consistently and coherently. Moreover, because different teams work on different services, they may lack a common way of handling and logging exceptions.

- Harder to secure. Again, this is mostly because different teams might use different technologies, tools, and frameworks, which may ship with different types of vulnerabilities. A good CI/CD factory helps alleviate some of these issues by statically scanning your code, as well as the open source libraries you rely on.

- Integration in the existing landscape is made harder because most enterprises have a broad set of legacy systems that cannot be migrated to microservices in the blink of an eye. This may force microservice designers to build anti-corruption layers to make their bounded context immune to what is happening in the legacy systems.

As you can see, microservices are complex and you should certainly not default to them. You should resort to microservices for complex domains and fast-moving businesses. On the other hand, should you build distributed monoliths instead of true microservices, the resulting complexity will still be higher, but you will automatically gain agility, scalability, and resilience. The bottom line is that a distributed monolith is more complex but still better than a monolith in terms of non-functional requirements. You will maximize these changes to have a system that is *fit for use*. Now, let's find out what hosting options we have for microservices.

Hosting microservices

Microservices are cloud native *par excellence*. However, you can host a microservice architecture on-premises because although it might sound contradictory, cloud native does not require cloud infrastructure. We will come back to this in *Chapter 6, Impact of the Cloud on the Software Architecture Practice*. Most true microservice architectures are hosted on containerization platforms. This is not an absolute requirement but it will make your life easier. Here are a few reasons why **Kubernetes (K8s)** is a good fit for microservices:

- **Scalability**: In K8s, every microservice corresponds to a deployment. Every deployment can specify its own scaling needs out of the box.

- **Resilience**: As we mentioned previously, using microservices increases the application's resilience because the failure of a single service should not affect all the others. With K8s, you can leverage self-healing, meaning that K8s will try to restart your failed containers.

- **Deployability**: Each microservice can be deployed independently using deployment techniques such as canary, blue/green, rolling updates, and so on.

- **Availability**: Because each microservice is separate from the others, we can easily identify mission-critical services and work with the **PodDisruptionBudget** resource type, which prevents unexpected outages, including during planned maintenance.

- **Support for polyglot architecture**: As we mentioned previously, microservices are polyglot because different teams can work with different technologies, providing the underlying hosting platform supports it. This is the case for container orchestrators.

To supplement K8s, you can also rely on service meshes such as **Linkerd, Istio**, and **Open Service Mesh** as they improve global oversight, ship with smarter load balancers (which are layer-7 protocol-aware), and come with **mTLS**, thus providing some sort of consistency across the different microservices.

Microservices in action

If you want to see microservices in action, I recommend that you look at the open source sample application at `https://github.com/dotnet-architecture/eShopOnContainers` or its variant at `https://github.com/dotnet-architecture/eShopOnDapr`, which simulates an online eShop. The second link also introduces **distributed application runtime (Dapr)**, which facilitates service discovery and communication with underlying data stores. Dapr is vendor-neutral and has connectors to dozens of systems. I am convinced that Dapr will be part of most microservice architectures in the next 5 years, but of course, I do not have a crystal ball.

Now, let's summarize this chapter.

Summary

I hope that you feel more comfortable now that you know about the three architecture styles we discussed in this chapter. As a software architect, you will often be able to choose either monolith or microservices. You will unlikely be entitled to go for SOA on your own because SOA is an enterprise-level effort, not a one man/woman show. However, you should have understood by now that going for non-SOA in an SOA-driven enterprise will be quite challenging.

Although architecture styles are high level, they have a significant impact, positive or negative, on software quality attributes.

In the next chapter, we are going to consider lower-level design choices and dive deeper into the code design patterns that are implemented by software developers.

5

Design Patterns and Clean Architecture

Although software architecture is not only about coding, a software architect is still required to have a vast knowledge about development in general and about design patterns in particular. In this chapter, we will explore some of the most frequent design patterns. We will be looking closely at code-related concerns.

We will more specifically cover the following topics:

- Understanding design patterns and their purpose
- Reviewing the **Gang of Four (GoF)**
- Delving into the most recurrent patterns and applying them to a use-case scenario
- Looking at clean architecture
- My top 10 code smells

This chapter should help you grasp the most recurrent design patterns and how to make use of them in your software architecture practice. These patterns must be well understood because they impact quality attributes such as modifiability, extensibility, reliability, maintainability, testability, scalability, and everything that relates to performance, as we have seen in *Chapter 3, Understanding ATAM and the Software Quality Attributes*.

Let's now review the technical requirements.

Technical requirements

To make some abstract concepts more concrete, some design patterns will be illustrated with .NET code samples. If you want to test the code locally, you will need **Visual Studio** or **Visual Studio Code** (**VS Code**). Both can be downloaded for free from Microsoft websites.

Note that the code is provided for illustration purposes only. All the code samples and diagrams are available at `https://github.com/PacktPublishing/Software-Architecture-for-Humans`.

Let's start with a definition and the rationale behind design patterns.

Understanding design patterns and their purpose

A **design pattern** is an admitted best practice to tackle a common problem. Many applications face the same challenges, as follows:

- They must be performant.

- They must be testable.

- They must be maintainable and should be able to grow over time.

- They must be portable to some extent.

- They must manage memory and the **central processing unit** (**CPU**) efficiently.

- They must be able to scale.

- They must support concurrency (thread safety).

The preceding list is only a subset of typical *cross-cutting concerns*. The purpose of a design pattern is to find the best approach to handle a common problem, regardless of whichever programming language is used.

Design patterns are inspirational and are not especially prescriptive about how the detailed implementation should be done. Design patterns remain high-level, and they aim to improve code quality as well as to ultimately adhere to the **SOLID** principles, a widely adopted set of principles crafted by Robert C. Martin.

SOLID stands for the following terms:

- **Single-responsibility principle** (**SRP**): This principle aims to master the scope of a given class and ensure a **separation of concerns** (**SoC**).

- **Open-closed principle (OCP)**: The idea behind this principle is to promote code extension while preventing changes to the current code. This can be achieved by using a certain level of abstraction through interfaces and abstract classes.

- **Liskov substitution principle (LSP)**: You can find plenty of definitions for this principle. One of the best ways to grasp it is through the rectangle-and-square example. As you know, in math, a square is also a rectangle. Let's say that we have a Rectangle base class with two separate SetWidth and SetHeight methods. Now, we implement a Square class that is a subclass of Rectangle, which seems to be a good fit since a square *is a* rectangle. But because we know a square must have an equal height and width, we override the base class methods to adjust the height whenever the width is set and vice versa. Doing so, the following construct will lead to an issue:

```
Rectangle rect = new Square();
rect.setWidth(10);
rect.setHeight(5);
Assert.Equal(50, CalculateArea(rect));
```

This is because when a square is passed, the resulting area will in this case be 25 instead of 50. In this case, we cannot substitute the rectangle with a square anymore because the square is unable to behave as a rectangle. More generally, the risk of breaking the LSP principle rises when the base and child classes have contradicting constraints. In this case, it might be better to simply split Square and Rectangle. The most recurrent, and often involuntary, observable manifestation of a broken LSP is the famous not implemented exception, which de facto prevents any substitution of the base class by a derived one.

- **Interface segregation principle (ISP)**: The purpose of this principle is to control the scope of an interface and make sure interface clients implement only what is necessary for their needs. You should therefore try to maximize the granularity of your interfaces.

- **Dependency inversion principle (DIP)**: The purpose of DIP is to make sure classes only depend on abstractions, not concrete classes. We will largely explore this principle later in this chapter.

> **Note**
>
> Design patterns already existed long before the creation of SOLID. They were therefore not created to adhere to SOLID, but because best practices are baked into these patterns, using them increases the chances of being SOLID-compliant by design.

The role of a software architect is to evaluate which patterns make more sense for a given situation. You should always try to keep the **Architecture Tradeoff Analysis Method (ATAM)** sensitivity and trade-off points in mind when choosing and applying these patterns in your development. You should not be dogmatic and blindly apply some patterns just for the sake of it. Doing so is called **cargo cult programming**, which consists of the ritual inclusion of patterns that serve no real purpose *in the context of your asset*. This is often a sign of a misunderstanding of these patterns, or, at best, it reflects an inadequate attitude.

A software architect is there to make sure every pattern is used in the interest of the asset and serves a real purpose. Software architects should try, more than anyone else, to build value for the business. One way to achieve this goal, in a timely and cost-efficient fashion, is to rely on well-proven existing solutions/frameworks whenever applicable. Crafting homemade frameworks and showing off how you can build complex things will make you virtually irreplaceable but not necessarily a good software architect. This is often the sign of an immature attitude. You must be in the skin of the business that pays for your work.

If you allow me a metaphor, you would not want to pay for a skyscraper if you initially asked for a house, even if the skyscraper was built with state-of-the-art and top-notch features. Similarly, you would not need the foundations of a skyscraper for a mere house. The message I want to convey here is that you should favor efficiency over effectiveness. Killing a fly with a hammer is *effective* but not *efficient*. This may sound obvious, but I have met countless developers who were very proud of themselves because they had built something that others did not understand well, and that is not a path you should take. Do not overengineer just for the sake of being proud of yourself! Let's now review the famous GoF.

Reviewing the GoF

The GoF originated from the unforgettable book, *Design Patterns: Elements of Reusable Object-Oriented Software*, published in 1994 and written by four authors— hence the name. This book popularized the notion of design patterns and proposed 23 of them. Many other patterns have been added since then. A good source to check most of the available patterns is this Wikipedia page: `https://en.wikipedia.org/wiki/Software_design_pattern`. Here, you can find the original patterns proposed by the GoF as well as more recently added ones. The first thing to know as a software architect is the different categories of design patterns, outlined as follows:

- **Creational**: Patterns in this category relate to the instantiation of objects. They mostly impact *performance* and *scalability*. They help prevent the waste of computing resources. Creational patterns, especially the **dependency injection (DI)** pattern, also improve the *testability* and *maintainability* of your code.

- **Structural**: Patterns in this category mostly focus on the composition of objects and classes. They mostly act upon the *readability* and *maintainability* of the code.

- **Behavioral**: The focus of behavioral design patterns is to handle the communication across objects. These patterns rule how objects interact with each other. Behavioral patterns have a strong impact on the *modifiability* of your code.

Note that I already made some links between the pattern categories and their impact on some quality attributes important in ATAM. If I had to rank these categories by priority during a code review or prior to developing anything, I would focus first on creational, then on behavioral, and finally on structural patterns.

The rationale is that, even if you write poor code in terms of readability and maintainability, this code might still do the expected job and be in line with **non-functional requirements (NFRs)** such as scalability and performance. In such a situation, the business is happy because the application is *fit for purpose* and *fit for use*. Conversely, if you write very nice code but you overlooked the performance and scalability aspects, the application might be *fit for purpose* but certainly not *fit for use*. Remember that the definition of quality is to satisfy both notions. I strongly encourage you to think about that whenever you review or are on the verge of designing a solution. Needless to say that a common objective among all design patterns is, of course, to increase reusability and aim for loosely coupled objects.

On its page, Wikipedia also proposes other categories, but there is no consensus about them. The preceding three categories are commonly adopted and understood. The Wikipedia page also covers some **architectural patterns**, which we talked about in the previous chapter. Let's now delve into the most recurrent patterns.

Delving into the most recurrent patterns and applying them to a use-case scenario

Admittedly, I have not performed any scientific research to assess which patterns are the most used ones. This is based on empirical observations in the field. In this section, we will focus on the most frequently used patterns, whichever type of application is built. In *Chapter 6, Impact of the Cloud on the Software Architecture Practices*, and *Chapter 7, Architectural Trends and Global Summary*, we will review patterns that are inherent to cloud-native and newer architectural trends, while the current section focuses on more traditional patterns.

The following diagram shows some of the patterns we are going to focus on:

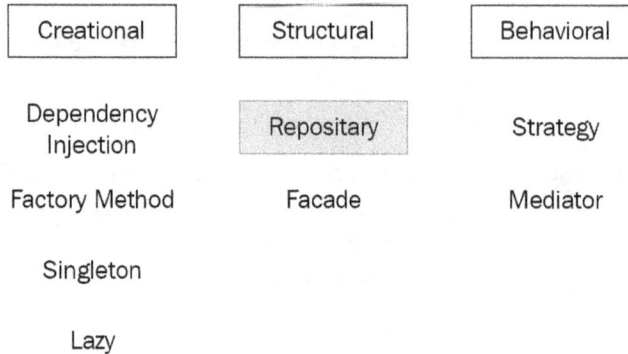

Creational	Structural	Behavioral
Dependency Injection	Repositary	Strategy
Factory Method	Facade	Mediator
Singleton		
Lazy		

Figure 5.1 – Design patterns we will focus on

In case you are wondering, the **repository** pattern has a dark background because its exact positioning is subject to debate. There is no real consensus about which category it belongs to. The pattern is itself often subject to controversy, but I will come back to that later in this section. I will not touch all the details of every pattern but rather try to highlight the essential parts and make sure you grasp what is key to remember.

Let's start with a pattern every software architect should undoubtedly know about – namely, DI.

Understanding the DI pattern

The purpose of DI, part of the creational category, is to decouple classes by linking through contracts, materialized by interfaces. Concrete classes implement the interfaces, and a concrete implementation of a given interface will be *injected* into the client class whenever that interface is encountered in the consumer class. There are multiple benefits to this approach, as outlined here:

- Testability is improved because we can easily replace a concrete implementation with a corresponding test class.

- The code is more maintainable because multiple concrete classes can implement the same interface differently, which increases the field of possibilities.

The following diagram illustrates DI:

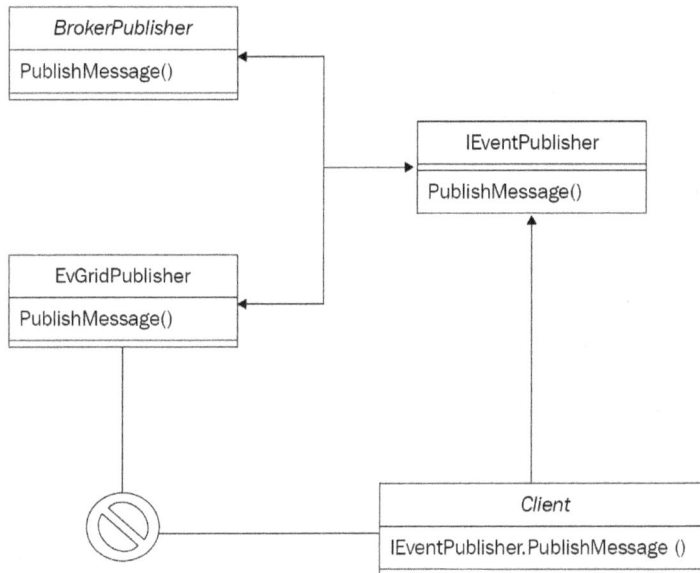

Figure 5.2 – DI diagram

Figure 5.2 shows that IEventPublisher is our contract—that is, our interface. Our Client class consumes this interface. BrokerPublisher and EvGridPublisher are both concrete implementations of IEventPublisher. The DI mechanism consists of injecting a concrete implementation of a given contract to a consumer class. This allows the consumer class to be unaware of the implementation details.

Figure 5.2 also shows that we could seamlessly inject BrokerPublisher or EvGridPublisher to the consumer. The only certainty the consumer has is that the injected concrete object (the dependency) implements the IEventPublisher contract. This allows for greater flexibility and extensibility. In the preceding example, we could extend our list of message publishers and have different implementations while not impacting the consumer side. This technique also greatly improves *testability* because we can easily inject a mocked object into a unit test. The injection can be achieved in the following two ways:

- **Direct**: This type of injection is typically done from a unit test.

- **Indirect**: This type of injection is done using a **DI container**. The role of the DI container is to resolve the concrete implementation of a given contract whenever an instance of that contract is encountered by the consumer class. Outside unit tests, this type of injection is the preferred way to go.

Let's see some code samples that illustrate the two modes. They are written in .NET, meaning that the specific implementation details may vary compared to another language. Even if you do not know .NET, just try to grasp the underlying concepts.

DI code – indirect mode

Indirect injection is provided by a DI container. There are numerous container frameworks available. By default, ASP.NET Core comes with its own default implementation. ASP.NET Core's startup class, and—more specifically—the `ConfigureServices` method, is where most of the DI plumbing happens. Here is an extract of that class:

```
public void ConfigureServices(IServiceCollection services){
    services.AddControllers();
    services.AddScoped<IEventPublisher, BrokerPublisher>();
}
```

Figure 5.3 – ConfigureServices

We can clearly see that the `IEventPublisher` contract is mapped to a concrete implementation represented by `BrokerPublisher`. In the case of ASP.NET Core, the built-in DI container is in charge of injecting consumers with a concrete implementation of the contract whenever an instance of a given contract is encountered. Here is our contract:

```
public interface IEventPublisher{
    4 references
    Task PublishMessage(string message);
}
```

Figure 5.4 – IEventPublisher contract

Our `BrokerPublisher` class is shown here:

```
public class BrokerPublisher : IEventPublisher{
    4 references
    public async Task PublishMessage(string message)
    {
        await Task.Run(() => {

            //SendMessageToBus()
        });
    }
}
```

Figure 5.5 – BrokerPublisher class

We pretend to send a message to a bus in the actual implementation of the SendMessage method, which is part of our contract. Similarly, another implementation of IEventPublisher could be this:

```
public class EvGridPublisher: IEventPublisher{
    4 references
    public async Task PublishMessage(string message){
        await Task.Run(() => {
            //SendMessageToGrid()
        });
    }
}
```

Figure 5.6 – Another implementation of IEventPublisher

This time, we pretend to send a message to an event manager. The two implementations show how flexible it is to work with interfaces. Now, in our **application programming interface (API)** controller, we let the DI container inject the relevant mapped concrete classes, as follows:

```
[ApiController]
[Route("[controller]")]
3 references
public class DemoController : ControllerBase{

    private readonly IEventPublisher _evPublisher;

    2 references
    public DemoController(IEventPublisher evPublisher){
        _evPublisher = evPublisher;
    }

    [HttpPost]
    2 references
    public async Task<IActionResult> PublishMessage(string message){
        await _evPublisher.PublishMessage(message);
        return new AcceptedResult();
    }
}
```

Figure 5.7 – DI container injecting classes

The constructor gets an instance of IEventPublisher as an argument, which in this case is replaced by an instance of BrokerPublisher, as defined in the startup class. It is possible to have multiple concrete classes implementing the same contract at the same time and add them to collections. What is key to remember in this example is that the controller is unaware of the implementation details. It will receive a concrete instance of an object that implements a certain contract. This makes it easier to write unit tests, as we will see in our next section.

DI code – direct mode

One of the biggest advantages of DI is its positive impact on the *testability* quality attribute. Let's see two different ways of testing the controller shown in the previous section. Still in ASP.NET Core, the following method makes use of a mocking framework:

```
[Fact]
0 references
public async Task PublishMessageTest()
{
    var _mockEventPub = new Mock<IEventPublisher>();
    var _demoController = new DemoController(_mockEventPub.Object);
    var actionResult = await _demoController.PublishMessage("test") ;
    Assert.IsType<AcceptedResult>(actionResult);
}
```

Figure 5.8 – Mocking framework

As you can see, we directly instantiate DemoController with a mocked representation of our IEventPublisher contract. This allows us to inject a fake object instead of the actual BrokerPublisher class we used previously. The reason you want to do this is that you do not want the actual actions to be taken during a unit test. Note that the usage of a mocking framework is not required, but it just makes your life easier. Here is an alternative implementation:

```
[Fact]
0 references
public async Task PublishMessageTestWithBasicDirectInjection(){
    var _demoController = new DemoController(new TestPublisher());
    var actionResult = await _demoController.PublishMessage("test");
    Assert.IsType<AcceptedResult>(actionResult);
}
```

Figure 5.9 – Alternative mocking framework

This time, we inject an instance of `TestPublisher`, whose implementation is shown here:

```
public class TestPublisher : IEventPublisher{
    4 references
    public async Task PublishMessage(string message)
    {
        await Task.Run(() => {
            //test use case()
        });
    }
}
```

Figure 5.10 – TestPublisher instance

This is just another variant of `IEventPublisher`, and this is where we can simulate whatever we want, thanks to the DI pattern. As you can see, in the preceding example, we performed a *direct* injection of the dependency via the constructor injection technique.

DI pattern wrap-up

Although we have seen a certain type of implementation, remember the following:

- The detailed implementation of DI is not prescriptive. Every implementation is OK as long as you make sure that an object gets its dependencies through one of the injection techniques.

- We only touched on constructor injection, but other possibilities exist, such as method injection and property injection, and these offer more granularity than constructor injection.

- DI plays a key role in abstracting away concrete classes from their consumers. You should, however, keep the number of abstraction layers low to avoid extra complexity and a negative impact on the *readability*.

- We used a mocking framework in one of our tests, but it is entirely up to you to see what you want to work with.

Usually, there are pros and cons for everything we do, but this does not apply to DI. It is a no-brainer that DI is strongly encouraged in every project. For this one, I authorize you to be the cargo cult developer. Let's now explore another well-known (anti-)pattern— namely, the singleton pattern.

Exploring the singleton design pattern

The singleton pattern, part of the creational category, is often subject to controversy because it is not always used appropriately. It is even considered an anti-pattern by many. However, this pattern is sometimes necessary for heavy objects. By heavy, I mean objects that take a long time to initialize or consume a lot of resources when being initialized.

The purpose of the singleton pattern is to instantiate a given object only once per application domain. This means that the object will be created once and shared for the entire lifetime of the application, which is why it is important to use it wisely because it also comes with a whole bunch of drawbacks. The following diagram shows a single instance that is shared across consumers:

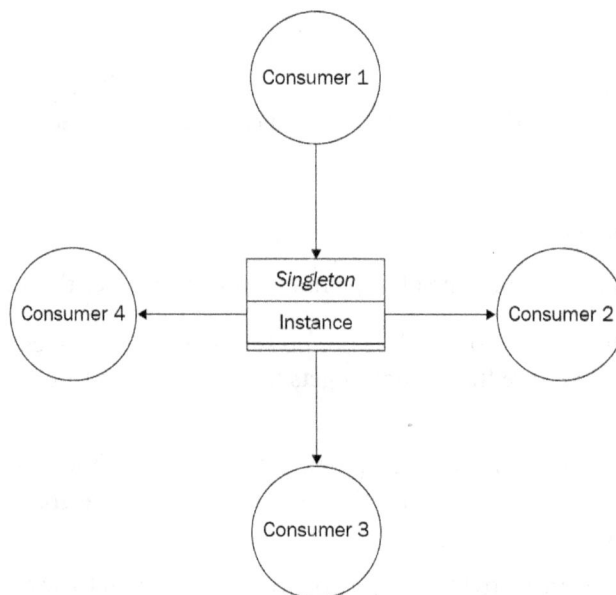

Figure 5.11 – Singleton design pattern

In the case of a web application, every separate **HyperText Transfer Protocol (HTTP)** request hits the same instance of a given singleton. Let's now see the singleton pattern in action.

Singleton pattern in action

The following code is the default startup method of a console application:

```
class Program{
      0 references
    static void Main(string[] args){
        Parallel.For(0, 10, i =>{
                Console.WriteLine("Iteration {0} - Identifier {1}",i,
                    ThreadSafeSingletonExample.instance.identitier);
                Console.WriteLine("Iteration {0} - Identifier {1}", i,
                    NotThreadSafeSingletonExample.instance.identitier);
            });

        Console.Read();
    }
}
```

Figure 5.12 – Singleton main method

As you can see from the preceding code snippet, I use a parallel for loop to simulate a high concurrency. In the loop, I get an instance of ThreadSafeSingletonExample and NotThreadSafeSingletonExample objects. Here is the implementation of NotThreadSafeSingletonExample:

```
class NotThreadSafeSingletonExample{
    static NotThreadSafeSingletonExample _instance = null;
    static Guid _id = Guid.Empty;

    1 reference
    private NotThreadSafeSingletonExample(){
        Thread.Sleep(100);
        _id = Guid.NewGuid();
    }
    1 reference
    public static NotThreadSafeSingletonExample instance{
        get{
            if (_instance == null)
                _instance = new NotThreadSafeSingletonExample()
            return _instance;
        }
    }
    1 reference
    public string identitier{get{return _id.ToString();}}
}
```

Figure 5.13 – Non-thread-safe singleton implementation

As their names indicate, one implementation is thread-safe while the other is not. With this example, I want to illustrate one of the major drawbacks of singletons—namely, thread safety. Under a high load, you may end up with multiple instances of a singleton if you do not pay attention to concurrency, which leads to unexpected outcomes because the main purpose of a singleton is to have a single instance in all circumstances. Moreover, such beginner errors may not be visible directly and might show up later once an application is already in production. This is not that easy to troubleshoot, so you'd better flush such issues out soon enough.

Coming back to the example, let's first start with the non-thread-safe implementation, the `NotThreadSafeSingletonExample` class. First, we see that `_instance` is a private class member, which is set when the private constructor is kicked off by the public `instance` property. In the `get` accessor, we set the value of `_instance` if it is `null`, and then we return it.

I have added a `sleep` statement to simulate a slow-initializing object. This code is supposed to create a single instance in all circumstances. The problem is that multiple threads can access the same resource at the same time. Under a race condition, when there is no synchronization mechanism in place, multiple threads could resolve the `if` statement to `true` at the same time and create an instance that results in breaking the concept of a singleton. A singleton is, by the way, not the only thing that is subject to concurrency issues. Every built-in collection may be subject to thread-safety issues as well.

To prevent concurrency issues, you must ensure multiple threads cannot create an instance at the same time. The way you achieve this is often specific to the programming language you work with. Some manual locking techniques are possible (single locking and double locking) but they come with their own caveats, such as having a negative impact on performance. In C# (specific to C# here), there is an easy way to overcome this issue, as shown in the following example:

```
class ThreadSafeSingletonExample{
    static readonly ThreadSafeSingletonExample _instance =
        new ThreadSafeSingletonExample();
    static Guid _id = Guid.Empty;
    0 references
    static ThreadSafeSingletonExample(){
        _id = Guid.NewGuid();
        Thread.Sleep(100);
    }
    1 reference
    public static ThreadSafeSingletonExample instance{
        get{
            return _instance;
        }
    }
    1 reference
    public string identitier { get { return _id.ToString(); } }
}
```

Figure 5.14 – Thread-safe singleton implementation

At first glance, this code looks very similar to that shown in *Figure 5.13*, except for one noticeable difference: the constructor is not only private, but it has also become **static**. C# ensures that static constructors are only called once per application domain. This small change makes a hell of a difference.

The following screenshot shows the output of the console program when executed:

```
Iteration 0 - Identifier fc71afb7-604d-4cbd-9b20-0c604565ed9a
Iteration 2 - Identifier fc71afb7-604d-4cbd-9b20-0c604565ed9a
Iteration 4 - Identifier fc71afb7-604d-4cbd-9b20-0c604565ed9a
Iteration 6 - Identifier fc71afb7-604d-4cbd-9b20-0c604565ed9a
Iteration 8 - Identifier fc71afb7-604d-4cbd-9b20-0c604565ed9a
Iteration 8 - Identifier 60fe4df0-a1da-4d90-806e-703dc9095f0f
Iteration 4 - Identifier 60fe4df0-a1da-4d90-806e-703dc9095f0f
Iteration 6 - Identifier 60fe4df0-a1da-4d90-806e-703dc9095f0f
Iteration 2 - Identifier 60fe4df0-a1da-4d90-806e-703dc9095f0f
Iteration 0 - Identifier f6189dd1-e279-45cc-a8eb-20b7de562d59
Iteration 1 - Identifier fc71afb7-604d-4cbd-9b20-0c604565ed9a
Iteration 1 - Identifier f6189dd1-e279-45cc-a8eb-20b7de562d59
Iteration 3 - Identifier fc71afb7-604d-4cbd-9b20-0c604565ed9a
Iteration 7 - Identifier fc71afb7-604d-4cbd-9b20-0c604565ed9a
Iteration 7 - Identifier f6189dd1-e279-45cc-a8eb-20b7de562d59
Iteration 3 - Identifier f6189dd1-e279-45cc-a8eb-20b7de562d59
Iteration 9 - Identifier fc71afb7-604d-4cbd-9b20-0c604565ed9a
Iteration 9 - Identifier f6189dd1-e279-45cc-a8eb-20b7de562d59
Iteration 5 - Identifier fc71afb7-604d-4cbd-9b20-0c604565ed9a
Iteration 5 - Identifier f6189dd1-e279-45cc-a8eb-20b7de562d59
```

Figure 5.15 – Singleton and concurrency

You can count if you do not believe it, but there are exactly *10 lines* reporting the `fc71afb7-604d-4cbd-9b20-0c604565ed9a` **globally unique identifier** (**GUID**), which corresponds to our thread-safe implementation. For the non-thread-safe implementation, we count two different GUIDs: one starting with `60fe4df0` and another one starting with `f6189dd1`. This means that we indeed got two different instances out of our poor implementation.

Singleton pattern wrap-up

The singleton pattern comes with the following benefits and drawbacks:

- Enhanced performance, especially for slow-initializing and heavy objects, which is a valid use case.

- Reduced usage of memory, because a single instance will of course consume less memory than multiple instances of the same object.

- Thread safety is not guaranteed by default, as we saw in the previous section.

- It can be harder to react to ecosystem changes such as underlying resource exhaustion, security token expiration, loss of network connectivity, and so on.

- It violates the SRP because it deals with both its own instantiation and some business logic.

Use the singleton pattern with caution and only when required. Valid use cases include cross-cutting concerns such as loggers, **object-relational models** (**ORMs**) when they remain thread-safe, third-party factories, and so on. Remember that singleton drawbacks are as important as their benefits and you should always double-check the concurrency aspects, as well as how you can detect/react to ecosystem changes. When working with rich frameworks such as .NET, you can also rely on the built-in DI container to assign a singleton behavior to a class that is not itself developed as such, while ensuring that only one single instance will ever be created.

Factory method

The factory method pattern, part of the creational category, is also part of a larger factory family of patterns such as **static factory** and **abstract factory**. Factory patterns, in general, are somewhat similar in their purpose to DI: decoupling concrete classes from client consumer classes. There is sometimes a bit of confusion between DI and factory patterns, precisely because they are both creational and pursue the same goals.

However, there's a noticeable difference between DI and factory patterns. On the one hand, with factory patterns, the consumer class is still triggering the creation process itself, even though it is delegated to other classes. On the other hand, with DI, client classes benefit from the **inversion of control (IoC)** principle and receive concrete implementations on the fly.

The following diagram illustrates the factory method pattern:

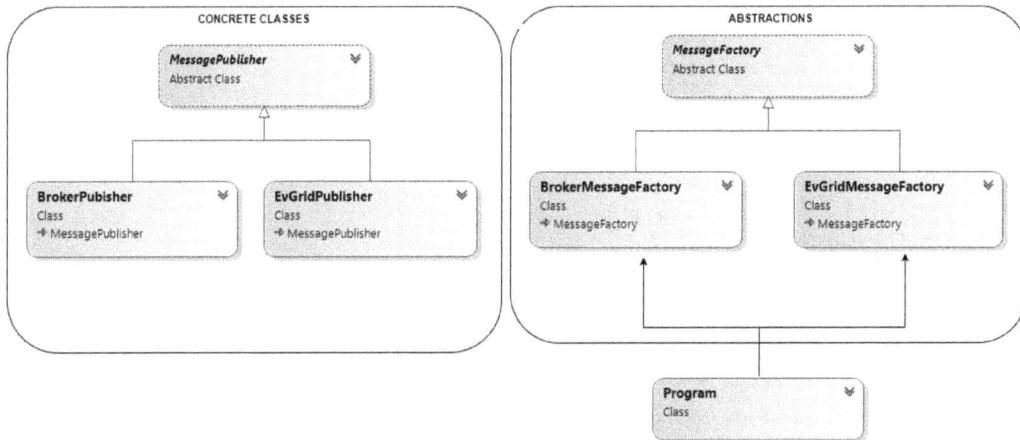

Figure 5.16 – Factory method pattern

On the left side of *Figure 5.16*, you can see concrete implementations of `MessagePublisher`, which is nothing other than our contract from the previous example, with the form of an abstract class instead of an interface.

Let's see the factory method in action.

Factory method in action

To illustrate the factory method pattern, we will reuse the message publishers of the previous section. Remember that we had two different concrete implementations that were implementing the same contract in a different way. This time, our contract is represented by the following abstract class:

```
public abstract class MessagePublisher{
    4 references
    public abstract string ProviderName { get; }
    2 references
    public abstract Task PublishMessage(string message);
}
```

Figure 5.17 – The factory contract

This abstract class declares two abstract members: `PublishMessage` and `ProviderName`. Here are the concrete implementations of that abstract class:

```
public class BrokerPubisher : MessagePublisher
{
    4 references
    public override string ProviderName => "BusMessage";
    2 references
    public override async Task PublishMessage(string message){
        await Task.Run(() => {
            //SendMessageToBus()
        });
    }
}
1 reference
public class EvGridPublisher : MessagePublisher{
    4 references
    public override string ProviderName => "EvGridMessage";
    2 references
    public override async Task PublishMessage(string message)
    {
        await Task.Run(() => {
            //SendToGrid()
        });
    }
}
```

Figure 5.18 – Concrete implementations of the MessagePublisher class

This is very similar to what we did in the DI example. The difference lies in how these concrete classes get indirectly used by the consumer class. This is done by declaring a few extra classes. The first one is the abstract creator class, which is illustrated here:

```
public abstract class MessageFactory{
    4 references
    public abstract MessagePublisher GetPublisher();
}
```

Figure 5.19 – Factory abstract creator class

The `MessageFactory` abstract class forces any inheriting class to implement the `GetPublisher` method, which returns a type of `MessagePublisher`, our contract. `GetPublisher` is our **factory method**. Now come the concrete creator classes that implement `MessageFactory`, as illustrated here:

```
public class BrokerMessageFactory : MessageFactory{
    4 references
    public override MessagePublisher GetPublisher(){
        return new BrokerPubisher();
    }
}
1 reference
public class EvGridMessageFactory : MessageFactory{
    4 references
    public override MessagePublisher GetPublisher(){
        return new EvGridPublisher();
    }
}
```

Figure 5.20 – Concrete creator classes

As you might have guessed, these concrete creator classes are the ones used by our consumer class, which in this case is a console program, as illustrated here:

```
static void Main(string[] args){
    Console.Write("Bus (1) or Event Grid (2): ");
    var choice = Console.ReadLine().Trim();
    switch(choice){
        case "1":
            Console.WriteLine(new BrokerMessageFactory().GetPublisher().ProviderName);
            break;
        case "2":
            Console.WriteLine(new EvGridMessageFactory().GetPublisher().ProviderName);
            break;
        default:
            Console.WriteLine("Oops, invalid choice");
            break;
    }
}
```

Figure 5.21 – Factory consumer program

We prompt the user to choose between options `1` and `2`. Option `1` causes the client class to get an instance of `MessagePublisher` using `BrokerMessageFactory`. The following screenshot shows the output of the program when executed:

```
Microsoft Visual Studio Debug Console
Bus (1) or Event Grid (2): 1
BrokerMessage

C:\Users\steph\Source\Repos\packt-software-architecture\patterns-abstractfactory\bin\Debug\netcoreapp3.1\patterns-factor
ymethod.exe (process 21888) exited with code 0.
To automatically close the console when debugging stops, enable Tools->Options->Debugging->Automatically close the conso
le when debugging stops.
Press any key to close this window . . .
```

Figure 5.22 – Factory method program output

Because we took option 1, the returned concrete implementation of the `MessagePublisher` contract is an instance of `BrokerMessage`. Let's now wrap up the section.

Factory method wrap-up

In the previous example, the benefit of using the factory method pattern was that the client (in this case, the console program itself) is not aware of the implementation details of the message publishers; it only needs to specify the factory to use. The client is immune from the changes happening in the creational process. Also, should a new message publisher type be needed, the only thing the client needs to do is to start using it.

The factory family patterns have the following advantages and drawbacks:

- They help decouple consumer and concrete classes.
- They help delegate the creation of complex objects to other classes.
- They increase complexity. As you probably noticed, the factory method example looks rather more complex than the DI one, and yet it is somewhat simpler than an abstract factory pattern. Do not abuse factory patterns.
- Factories can be combined with DI.

Let's now discuss our last creational pattern—namely, the lazy loading/initialization pattern.

Lazy loading/initialization pattern

The lazy loading/initialization pattern (both names are accepted) is pretty easy and very handy. As the name indicates, it will lazily initialize an object of a given type only when an explicit call is performed by the code. The purpose of the lazy pattern is to avoid a waste of resources and defer these until the application really requires an instance of a given object to function.

Lazy loading in action

Here is a very simple example of the lazy loading pattern in a C# console program:

```
static void Main(string[] args){
    Stopwatch sw = new Stopwatch();
    sw.Start();
    //does not cause a delay
    Lazy<DemoClass> _lazy = new Lazy<DemoClass>();
    Console.WriteLine(sw.ElapsedMilliseconds);
    //calling the object property causes its lazy loading
    Console.WriteLine(_lazy.Value.DemoProperty);
    Console.WriteLine(sw.ElapsedMilliseconds);
    Console.Read();
}

3 references
public class DemoClass{
    public readonly string DemoProperty = "a value";
    0 references
    public DemoClass(){
        Thread.Sleep(5000);
    }
}
```

Figure 5.23 – Lazy loading in action

In our `DemoClass` object, we simulate a slow initialization by adding a `Thread.Sleep` statement. In the `Main` method of the console program, we make use of a `Stopwatch` object to calculate the elapsed time between the different instructions. The output of the preceding program looks like this:

Microsoft Visual Studio Debug Console

```
0
a value
5060
```

Figure 5.24 – Lazy loading in action (continued)

As you can see from *Figure 5.24*, no time elapsed between the start of our `Stopwatch` object and the creation of our `lazy` object. The lazy behavior is ensured by the `Lazy` keyword, which is a built-in C# instruction. However, when calling explicitly the property of our `DemoClass` object through `_lazy.Value.DemoProperty`, we can see that our second printout of the elapsed time shows `5060` milliseconds. This proves that the actual initialization of the `DemoClass` object was delayed until we called one of its properties (it is the same with methods, of course). The same program without the lazy pattern looks like this:

```
static void Main(string[] args){
    Stopwatch sw = new Stopwatch();
    sw.Start();
    DemoClass _notLazy = new DemoClass();
    Console.WriteLine(sw.ElapsedMilliseconds);
    Console.Read();
}
```

Figure 5.25 – Non-lazy initialization

Now, we instantiate the same `DemoClass` object the normal way. The following screenshot shows the outcome:

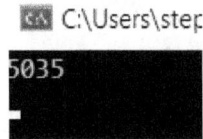

Figure 5.26 – Non-lazy DemoClass object

Lazy loading wrap-up

The direct instantiation causes the program to take about 5 seconds before it prompts the user for a key entry. Because the lazy pattern is very simple, there is no real need to add a wrap-up section, but let's still see some of its benefits and drawbacks, as follows:

- The major benefit of the lazy pattern is to prevent a waste of resources when the concrete instances of a given object are not needed.

- This major benefit might also be its main drawback. If you were to use lazy loading for everything, you would not have a clear picture of the actual resources your application needs to fully function. Make sure to evaluate this when testing the application.

Let's now go through some of the behavioral patterns.

Strategy pattern

The strategy pattern, part of the behavioral category, is used to let consumer classes decide on which algorithm to choose from a family of related algorithms. The purpose of this pattern is to avoid `if-else` or `switch` constructs directly in the client, to handle different implementations, and let it pick the right strategy instead. Here is a class diagram of the sample that comes next:

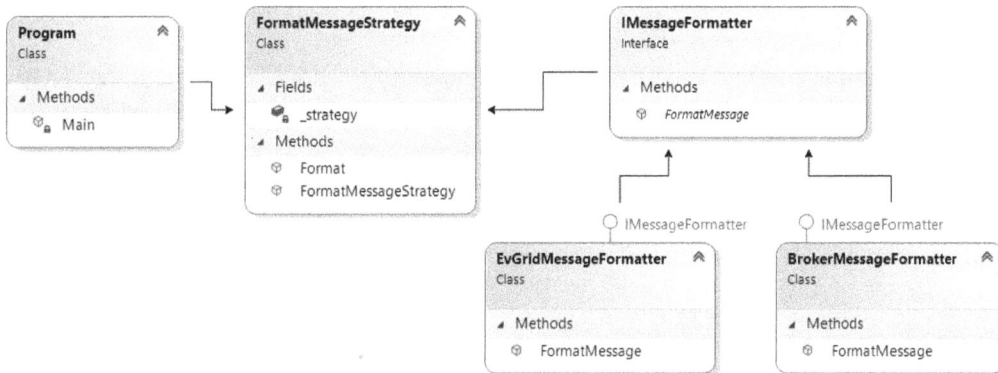

Figure 5.27 – Strategy pattern

The `Program` class is our consumer. `FormatMessageStrategy` is our **context** class, which lets the consumer specify at runtime which strategy to use. `IMessageFormatter` is our strategy contract, while `EvGridMessageFormatter` and `BrokerMessageFormatter` are different implementations of that strategy. This is comparable to what we have seen before, but I added some validation bits to the picture. Here, we have two different concrete strategies to format messages according to their target recipient. Let's see the strategy pattern in code.

Strategy pattern in action

The strategy pattern helps prevent the following code constructs in the consumer class:

```
if (msg.MessageType == Message.MessageTypes.Bus){
    //write code to format bus messages
}
else if (msg.MessageType == Message.MessageTypes.Grid){
    //write code to format grid messages
}
```

Figure 5.28 – Strategy pattern prevents inline decisions

Here, you would add some formatting logic directly into the client. The goal of the strategy pattern is instead to delegate this logic to dedicated strategies and choose the appropriate one at runtime. Here is the code of the strategy contract:

```
public interface IMessageFormatter{
    3 references
    string FormatMessage(string message);
}
```

Figure 5.29 – Strategy contract

This is a very simple contract that specifies that every strategy should implement the FormatMessage method. Here, we can see two strategies that implement our contract:

```
public class EvGridMessageFormatter : IMessageFormatter{
    3 references
    public string FormatMessage(string message){
        Console.WriteLine("in EvGridMessageFormatter");
        //let's pretend we formatted the message for evengrid
        return message;
    }
}
1 reference
public class BrokerMessageFormatter : IMessageFormatter{
    3 references
    public string FormatMessage(string message){
        Console.WriteLine("in BrokerMessageFormatter");
        //let's pretend we formatted the message for a bus
        return message;
    }
}
```

Figure 5.30 – Concrete strategies

We pretend to format the message differently according to the strategy. Now comes our context class, the one that is used by our consumer class, as follows:

```
public class FormatMessageStrategy{
    private IMessageFormatter _strategy;
    2 references
    public FormatMessageStrategy(IMessageFormatter strategy){
        _strategy = strategy;
    }
    2 references
    public string Format(string message){
        return _strategy.FormatMessage(message);
    }
}
```

Figure 5.31 – Context class

The context class has a private member of the `IMessageFormatter` type. Its constructor takes a strategy as input and assigns it to the private member. This lets the consumer class specify which strategy to use at runtime. The `Format` method can be called by the consumer class and, in turn, calls the `FormatMessage` method of the strategy that was passed in. Now, the next code block shows how that context class can be used from within the consumer class:

```
static async Task Main(string[] args){
    IEventPublisher brokerPublisher = new BrokerPublisher();
    await brokerPublisher.PublishMessage(
        new FormatMessageStrategy(new BrokerMessageFormatter()).Format("a message"));
    IEventPublisher gridPublisher = new EvGridPublisher();
    await gridPublisher.PublishMessage(
        new FormatMessageStrategy(new EvGridMessageFormatter()).Format("a message"));
    Console.Read();
}
```

Figure 5.32 – Strategy consumer code

We subsequently call both strategies. The following screenshot shows the output of the console program when executed:

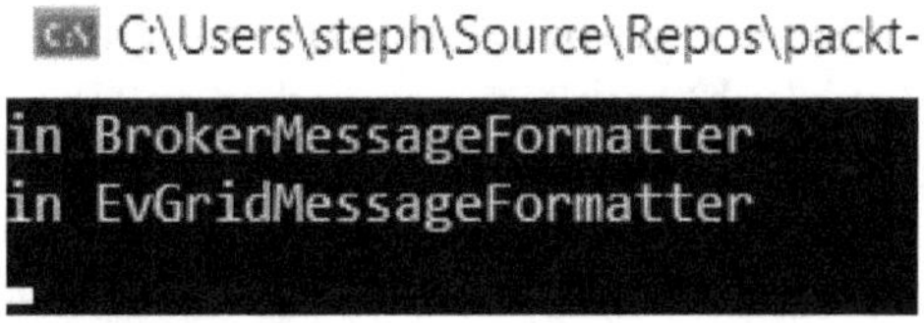

Figure 5.33 – Strategy pattern console output

Strategy pattern wrap-up

You might be wondering how the DI method differs from the strategy pattern, and you would be right to do so. Both patterns enable us to pass concrete objects to client classes through method injection. A big difference between the strategy pattern and DI is the category to which they belong. Remember that DI is a part of the creational patterns and, as such, they enter into play at the time objects get created.

The strategy pattern is a part of the behavioral category, and as such, it is intended to implement multiple behaviors to react to application events, such as user inputs or button clicks. When you use the strategy pattern, it is systematically with the intention of implementing multiple concrete classes of a given contract, while this is not especially the case (other than in unit tests) with DI only.

Another difference lies in the fact that with the strategy pattern, the consumer class must be aware of the concrete strategy classes, while DI only requires knowledge of the contract.

To wrap up the strategy pattern, let's review some of its benefits and drawbacks, as follows:

- It helps you delegate the business logic to strategies, instead of writing it directly in client classes. This satisfies the single responsibility principle of SOLID.

- Should the business logic change for a given strategy, changes will be automatically reflected in the client consumer classes.

- It is fairly easy to get started and improves readability, maintainability, and extensibility.

- Unlike DI, the client needs to be aware of the concrete strategies, which is not ideal.

Let's now explore the mediator pattern.

Mediator pattern

The mediator pattern, part of the behavioral category, facilitates the communication between objects. You can see it as a man in the middle, as a *dispatcher*. Typical examples are a chat room that dispatches messages between senders and receivers, a notification engine, **publish/subscribe (pub/sub)** architectures, and so on. The following diagram shows a depiction of the mediator pattern:

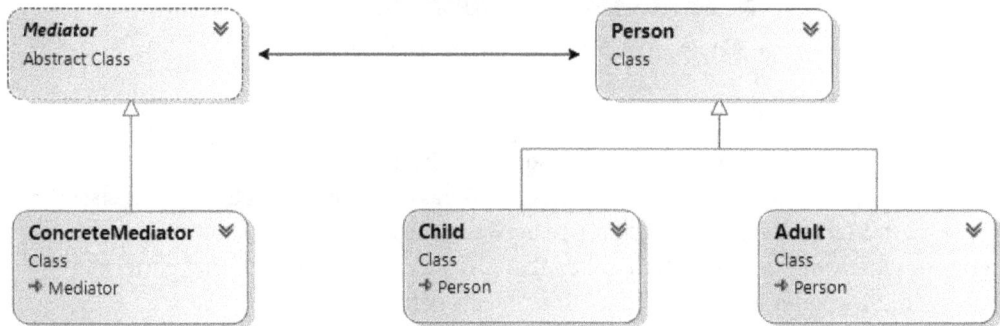

Figure 5.34 – Mediator pattern diagram

The Mediator class defines the communication contract and the ConcreteMediator class holds the communication logic, while all the other classes talk to each other through the concrete mediator.

Classes that participate in the communication all inherit from `Person`, which in turn holds a reference to the concrete mediator. Most examples that illustrate this design pattern look alike. I tried to innovate a little bit, but we will see another example in *Chapter 7*, *Architectural Trends and Global Summary*, about using the mediator pattern together with **Command Query Request Segregation (CQRS)**.

Mediator in action

As always, we start by defining our contract, as follows:

```
interface IMediator{
    4 references
    public abstract void Register(Person p);
    2 references
    public abstract void Send(string from, string message, audience to);
}
```

Figure 5.35 – Mediator contract

The contract is an interface or an abstract class. We then implement the concrete mediator, as illustrated here:

```
class ConcreteMediator : IMediator{
    private Dictionary<string, Person> _persons =
        new Dictionary<string, Person>();
    4 references
    public void Register(Person p){
        _persons.TryAdd(p.Name, p);
        p.concreteMediator = this;
    }
    2 references
    public void Send(
        string from,string message, audience to){

        var persons = (audience.adult == to) ? _persons.Values.Where(
            p=>p.Name != from && p.GetType().Equals(typeof(Adult))) :
            _persons.Values.Where(p=>p.Name != from);
        if(persons.Count()>0)
            foreach (var person in persons)
                person.Receive(from, message);
    }
}
```

Figure 5.36 – Concrete mediator

In the `Register` method of the concrete mediator, we add the person to our dictionary or replace them. In the `Send` method, we handle the business logic to forward messages to the relevant recipients. Admittedly, the filtering logic is not robust since the name is not a good ID, but this is irrelevant for the pattern demonstration. We also pass the current concrete mediator to the `Person` object, as follows:

```
class Person{
    6 references
    public string Name { get; }
    2 references
    public Person(string name) => this.Name = name;
    2 references
    public ConcreteMediator concreteMediator { set; get; }
    2 references
    public void Send(string message, audience to = audience.everyone){
        concreteMediator.Send(Name,message, to);
    }
    1 reference
    public void Receive(
        string from, string message){
        Console.WriteLine("{0} - received {1}: from {2}",
            Name,message,from);
    }
}
```

Figure 5.37 – Person object

The important bits of the preceding code block are in the `Send` method, where the `Person` class calls back the concrete mediator to handle the sending of messages. The `Send` method also has an optional `audience` parameter, which allows us to filter out target recipients when sending messages. The following code shows two persons' flavors:

```
class Adult : Person{
    2 references
    public Adult(string name): base(name){}
}
2 references
class Child : Person{
    1 reference
    public Child(string name): base(name){}
}
```

Figure 5.38 – Person variants

Now, from the main program, we can start using our classes, as follows:

```
static void Main(string[] args){
    ConcreteMediator cm = new ConcreteMediator();
    Person a1 = new Adult("adult 1");
    Person a2 = new Adult("adult 2");
    Person c1 = new Child("child 1");
    cm.Register(a1);
    cm.Register(a2);
    cm.Register(c1);
    a1.Send("Hello adults",audience.adult);
    a1.Send("Hello everyone");
    Console.Read();
}
```

Figure 5.39 – Consumer code

Notice how we register all person instances to the concrete mediator. Here is the program's output when executed:

```
C:\Users\steph\Source\Repos\packt-software-architecture\patte
adult 2 - received Hello adults: from adult 1
adult 2 - received Hello everyone: from adult 1
child 1 - received Hello everyone: from adult 1
```

Figure 5.40 – Mediator pattern program output

Figure 5.40 shows that only `adult 2` received the message from `adult 1` because they explicitly targeted adults. Their second message is for everyone, hence the reason why `child 1` also received it.

Mediator pattern wrap-up

The mediator has some advantages and drawbacks, outlined as follows:

- It ensures one-to-one and one-to-many communication processes.
- It decouples the objects that communicate with each other.
- It isolates the communication logic in the concrete mediator.
- As a disadvantage, the SRP might be compromised should the concrete mediator implementation become complex over time.

Let's now tackle structural design patterns.

Facade design pattern

The facade pattern, part of the structural category, helps hide the complexity of concrete classes by exposing a simple and clean interface to consumer classes. It acts as a proxy between consumers and concrete implementations. We will come back to the facade design pattern in our final chapter, to see a code-free approach to apply it. In the meantime, let's see the repository design pattern, which is a data-specific facade implementation.

Repository design pattern

The repository design pattern is a data-specific implementation of the facade design pattern. You use it as a mechanism between your API controllers and your **data transfer object** (**DTO**) that is used by the ORM. The repository is an abstraction of a collection of objects. Although there is some controversy about the added value brought by the repository pattern, its main purpose is to decouple the data and the business-domain worlds. The debate about the usefulness of the repository pattern lies in the fact that it is sometimes considered redundant with the ORM.

In the .NET world, the main ORM is **Entity Framework**, which ships with built-in repositories. Adding your own repository to the mix is considered overkill by some and indispensable by others. I am not going to make a decision in this debate, but the main argument of using repositories is to improve testability and use them to hold the business logic. You can see a diagram of the repository pattern diagram here:

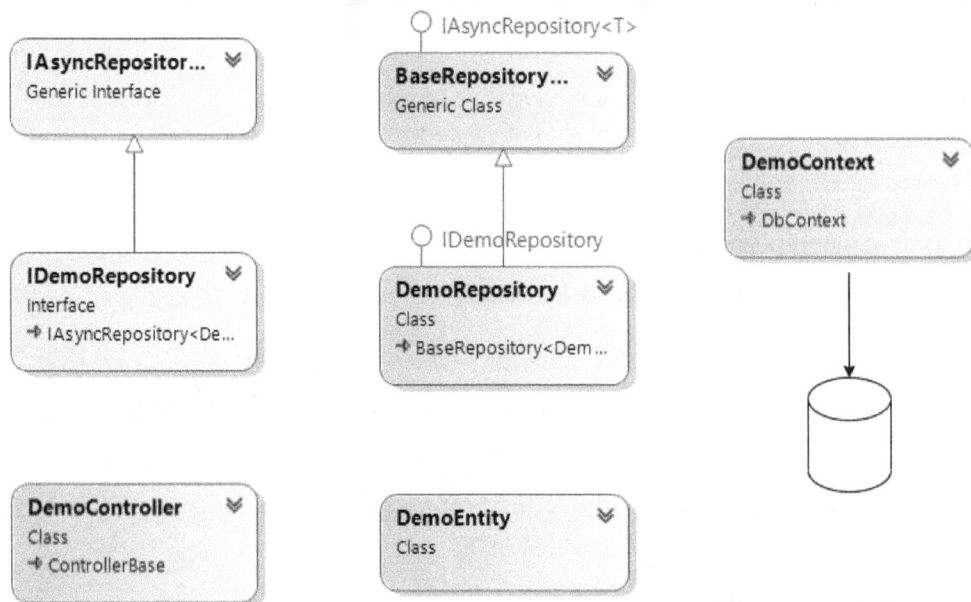

Figure 5.41 – Repository pattern diagram

On the left-hand side of *Figure 5.41*, you can see a generic repository represented by IAsyncRepository and a business-specific one represented by IDemoRepository. The DemoController class gets an instance of IDemoRepository through DI, for which the concrete class is DemoRepository, which in turn inherits from the BaseRepository class.

Both BaseRepository and DemoRepository leverage DemoContext (ORM) to interact with the underlying data store. Finally, DemoEntity is simply a representation of a data entity. Since we have already touched on DI and controllers before, I will only list the most important code blocks in our next section. You can find all the code on GitHub, as explained in the *Technical requirements* section of this chapter.

Repository pattern in action

Let's first start with our generic repository, as follows:

```
public interface IAsyncRepository<T> where T : class{
    0 references
    Task<T> GetByIdAsync(Guid id);
    0 references
    Task<IReadOnlyList<T>> ListAllAsync();
    0 references
    Task<T> AddAsync(T entity);
    0 references
    Task UpdateAsync(T entity);
    0 references
    Task DeleteAsync(T entity);
}
```

Figure 5.42 – Generic repository

The repository is generic because it takes an input of T, which is the .NET way to handle generics. All the methods declared in the interface are also generic because they do not serve any specific business purpose. They simply represent the typical **create, read, update, and delete (CRUD)** operations.

This generic repository is particularly the reason why the repository pattern is subject to controversy. If you only stick to this implementation, you simply add redundant code because, by default, ORMs also come with such default CRUD operations over collections of objects.

Some may argue that a generic repository also gives you an opportunity to abstract away the ORM, should you change it in the future. Such levels of abstraction are also recommended by *clean architecture*, which we will talk about in our next section. This is where you must exercise good judgment and identify trade-offs to make sure you do not overengineer the solution and that it makes sense in your own context.

But of course, there's more to it, and that is the specific business domain repository, whose contract is listed here:

```
public interface IDemoRepository : IAsyncRepository<DemoEntity>{
    2 references
    Task<IEnumerable<DemoEntity>> ListOnlyOddEntities();
}
```

Figure 5.43 – Domain-level repository

The reason why it is business-specific is that it specifies an extra non-generic method looking like a specific query. This type of repository starts to bring value because it holds business logic. Let's now see a truncated (for brevity) version of the `BaseRepository` class, as follows:

```
public class BaseRepository<T> : IAsyncRepository<T> where T : class{
    protected readonly DemoContext _dbContext;

    1 reference
    public BaseRepository(DemoContext dbContext) {
        _dbContext = dbContext;
    }
    2 references
    public async Task<IReadOnlyList<T>> ListAllAsync(){
        return await _dbContext.Set<T>().ToListAsync();
    }
    other methods
}
```

Figure 5.44 – Base repository class

Our `BaseRepository` class is also generic and can take any entity. It implements `IAsyncRepository<T>`, our generic repository contract. As you can see, it takes an instance of the `DemoContext` object, which is nothing other than our ORM's entry point. Now comes our tailor-made repository, as follows:

```
public class DemoRepository : BaseRepository<DemoEntity>, IDemoRepository
{
    0 references
    public DemoRepository(DemoContext dbContext) : base(dbContext) { }
    2 references
    public async Task<IEnumerable<DemoEntity>> ListOnlyOddEntities(){
        return await Task<IEnumerable<DemoEntity>>.Run(() => {
            return _dbContext.DemoEntities.ToList().Where((c, i) => i % 2 != 0)
                as IEnumerable<DemoEntity>;
        });
    }
}
```

Figure 5.45 – Tailor-made repository

DemoRepository derives from BaseRepository and passes in the DemoEntity type. It also implements IDemoRepository through the concrete implementation of the ListOnlyOddEntities method. If you know a little bit about .NET, you have probably identified a *code smell* here. The odd/even check is done in memory and causes the ORM to produce a SELECT * FROM ... statement, which is never good performance-wise. I did this on purpose to show you how easy it is to write very extendable code but leave such poor constructs in it. By the end of this chapter, I will let you know what I consider to be the most important things to catch in a code-review round.

Repository pattern wrap-up

Here are a few things to consider when using the repository pattern:

- Do not stick to the generic repository only. You should aim to encapsulate the business logic into more specific repositories.

- Do not abstract away the ORM, which is itself an abstraction of the underlying data store. Probably 99.99% of the time, you will only stick to a single ORM for the lifetime of the application.

- Repositories help improve testability.

- Keep the number of abstraction layers manageable. I'll leave it to you to evaluate what *manageable* means.

Let's now go through a small use case to see if you can apply what you have just learned about design patterns.

Design patterns use case

Let's go through the same scenario as the one we used for ATAM in *Chapter 3, Understanding ATAM and the Software Quality Attributes*, but this time, don't focus on the NFRs. Try to evaluate which code design patterns might be in scope for the following scenario:

> *Contoso needs to provide a data upload channel for its customers. Uploaded data files may contain errors and must go through a data-cleansing and data-wrangling phase. If errors cannot be automatically fixed, an error file should be returned to the sender through a callback notification. After this first data-check and transformation phase, the resulting validated data is routed to the relevant systems for further handling. In addition, customers require that data can only be processed in Europe for sovereignty reasons, and they are used to File Transfer Protocol (FTP) systems.*

> *The upload channel should always be available, while the actual handling of the data might be deferred in case of a system outage. It is expected by Contoso's customers to be notified back within 24 hours. The total volume of data uploaded by customers is about 1 terabyte (TB) per month and, during peak times, up to 250 customers may upload data at the same time. For regulatory reasons, the retention time of original files sent by the customers is 5 years.*

I did not change a single comma. To be honest, if I were confronted with this scenario, I would directly try to rely on a specialized data service for the data-cleansing and data-wrangling activities, but for the sake of this exercise, let's consider that Contoso does not have any of these tools. Contoso can only rely on a software architect and a few developers to implement the solution. Before reading the next paragraph, try to think about which patterns could play a role in that scenario. Limit yourself to the patterns discussed in this chapter.

OK—let's try to find answers. If we exclude the NFRs from the equation, here are a few elements that we can consider:

- The input data must be validated. Because there will be probably different types of validation, it might make sense to use validation strategies using the *strategy* pattern.

- Because every use case must deal with dependencies, you must of course consider using the *DI* pattern. As stated before, DI is a no-brainer and should always be used.

- Because you want to simplify the lives of Contoso customers, you might want to provide them with a single endpoint to upload files. If that is the case, you will probably rely on the *facade* design pattern. I have not elaborated much on this in this chapter, but the definition I made is enough to think it could be a useful pattern for this scenario.

- Many files must be processed. You should therefore consider handling them in an asynchronous way. A *mediator* or some variant could be used to decouple the receiving process from the handling process.

Let's see why I did not *yet* select the other patterns, as follows:

- **Repository**: It is not clear whether we only process data and route it or whether we need to persist anything in a data store. It might be necessary to persist some audit trail information, but this is not explicit yet.

- **Singleton, lazy, and factory**: At this stage, there is no indication yet that singletons or factories would be required or useful. We would need to investigate more deeply to decide.

Of course, we all know that the devil is in the detail, but if I had only one piece of advice to give you, it would be to approach things from a high level first before diving into in-depth implementation considerations. Let's wrap up the design patterns section.

Design patterns wrap-up

In this chapter, we only touched upon a few design patterns, but you might have noticed yourself that they all bring a certain level of abstraction to an application, to promote extensibility, maintainability, and testability. If you had to remember only one pattern, you should undoubtedly go for DI because this is, in my opinion, one of the most important concepts.

Note that while abstraction is good for extensibility, it may sometimes also bring its own caveats, such as increased complexity, should the number of abstraction layers become too high. Another aspect to consider when implementing patterns is the number of resulting objects that are created in the application because it may as well have a negative impact on system performance.

The message I want to convey here is that you should always balance extensibility, readability, maintainability, and testability with performance and efficiency. *Do not be the cargo cult developer.* Think, and do not obey blunt rules that may not be adequate for your context! Let's now look at clean architecture.

Looking at clean architecture

Beyond code design patterns themselves, you can also rely on more structural foundations. **Clean architecture** (`https://blog.cleancoder.com/uncle-bob/2012/08/13/the-clean-architecture.html`), proposed by Robert C. Martin, also known as *Uncle Bob*, questions the prevalence of frameworks and technical choices over the pursued business *intent*. Clean architecture regroups the best of **hexagonal architecture** (`https://en.wikipedia.org/wiki/Hexagonal_architecture_(software)`) and **onion architecture** (`https://jeffreypalermo.com/2008/07/the-onion-architecture-part-1/`). In essence, clean architecture decouples the business logic from all the rest and adds more layers to an application. The goal of clean architecture is to make the business layer immune from changes happening anywhere else in the application. The business layer should only change if there is a real business need to cover. A clean architecture ensures that the business layer remains in sustainable homeostasis. The following diagram shows the structure of a clean architecture (Uncle Bob's original drawing is available on his website at `https://blog.cleancoder.com/uncle-bob/2012/08/13/the-clean-architecture.html`):

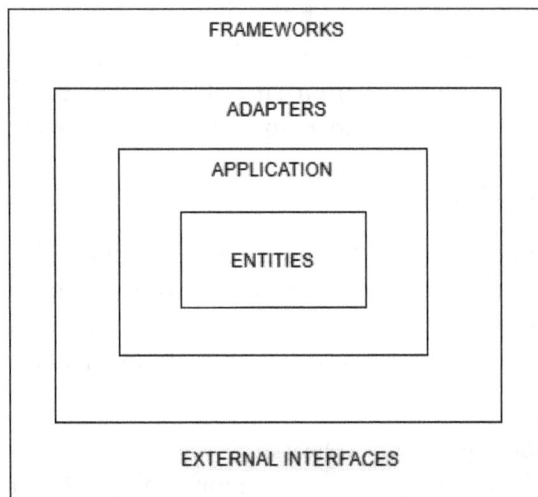

Figure 5.46 – Clean architecture

Only outer borders should know about inner borders, not the other way around. Inner borders host the core business logic, while outer borders are facilities and implementation details. When this rule is respected, inner borders cannot be impacted by a change occurring in the outer borders because they know nothing about them. Inter-border communication is mostly ensured by DI and IoC, which we have seen previously.

Conceptually, clean architecture is a *domain-centric software architecture*. This type of architecture grants a high (too high?) importance to the *fit-for-purpose* question. Frameworks, database engines, tooling, and the like are considered to be *implementation details*. However, while I understand the ambition, some push it too far by simply ignoring the so-called details and focusing *only* on the business logic.

Remember that the definition of quality is *fit for purpose* and *fit for use*. Some clean architecture aficionados consider that, in a house, the only thing that matters is space, usability, comfort, and so on, while the materials used for building it (bricks, tiles, and more) are just *implementation details*. I could not disagree more with that statement.

Uncle Bob himself states that *the database engine you work with is a detail* and that *the web itself is a detail*, and that basically everything that is not directly serving the business logic itself is a detail. While he is right from a business perspective, these *details* are our day-to-day reality, and we cannot get rid of them just like that. On the contrary—these details sometimes help boost productivity, and yes: we might consciously tightly couple ourselves to specific technical frameworks and tools because they also bring us something back. We evolve in a highly technical industry. If every tool, every framework, every database engine were only a detail, anyone could deal with them. You would not need to hire developers, **Information Technology** (IT) pros, and so on. Perhaps, the business folks would design and develop solutions themselves.

To come back to the house metaphor, in our *fit-for-use* quest, the house must be resistant to storms, heavy rain, humidity, and the like, else you might end up with a very comfortable house but be forced to move a few years later because it became inhabitable in the meantime. Of course, these factors depend on the context, *as always*. You will not invest in the most resistant tiles if you live in a region where it never rains. Similarly, you are unlikely to pay for specific skiing insurance if you never go skiing. Well, all of this seems to be common sense, but that is not especially what I notice in the field.

This trend toward domain-centric architectures and conceptual approaches such as **domain-driven design** (DDD) is certainly a good thing, provided you do your homework, which consists of conducting proper business analysis (together with business analysts and business architects) to grasp the business domain you are building a solution for. And even then, you should still not neglect the *details*.

These domain-centric architectures only prove their value if that prerequisite is respected, and they are only worthwhile with fast-changing business needs. I have seen countless times teams rushing to clean architecture and/or DDD, regardless of what they had to do, while not having a single clue of what really matters for their business. In such situations, these patterns are abused and tend to create confusion instead of bringing solutions to the fore. To be clear, we should not go from one extreme to the other. We should not start neglecting our technical foundations as we have sometimes neglected or limited the pursued business value by technicalities. By the way, we should not oppose technicalities and business value.

Let's remember that the digital natives took advantage of technology as an enabler for new business perspectives. What makes a developer and a software architect different from a business analyst are precisely these technicalities. Let's also remember that code is not the only thing to be considered. Cloud-native apps are a true example of this—they leverage containers (a pure infrastructure thing) and orchestrators to build resilient solutions by leveraging those system capabilities.

In one of his talks (`https://www.youtube.com/watch?v=yPvef9R3k-M`), Eric Evans, who came up with the DDD concept, admitted that microservices could represent the physical boundary of a bounded context. Here again, container platforms are heavily supporting microservice architectures, although they are not the only option. I am convinced that the ecosystem takes a very important place in software architecture, while the code you develop yourself tends to become tinier. Whether you should use clean architecture or not depends on the asset you build and on the ATAM round you had about that asset.

If you identify that some quality attributes such as usability, upgradability, and maintainability prevail over security, scalability, and so on, you may indeed find it appropriate to go for clean architecture, else you might as well consider it overkill for your use case. Let's now see what I consider important in a code-review exercise.

My top 10 code smells

A **code smell** is a code construct that could lead an application to crash or encounter unexpected issues. As I showed in the repository pattern section, it is very easy to leave rubbish inside well-implemented design patterns. Admittedly, most modern **continuous integration/continuous delivery (CI/CD)** factories ship with code-analysis tools such as *SonarQube* that analyze code and detect code smells automatically. Nevertheless, here are the top 10 things I consider when reviewing code or assessing the results produced by such a tool:

1. **Memory leaks will for sure lead to process crashes**: The sooner you detect them, the better.

2. **Improperly disposed objects**: While memory leaks are annoying, they are easy to troubleshoot. Conversely, disposed objects that are still used somewhere are harder to find and often lead to erratic behaviors.

3. **ORM usage**: It is great to abstract away the underlying data store but, in some situations (**input/output (I/O)**-intensive use cases, and so on), it is better to skip ORMs completely, or at least to double-check queries/commands that are produced. The poor code construct that I included in the repository sample is a good illustration of such a waste of resources.

4. **Security flaws**: In our API-driven world, it is not uncommon to find endpoints that are unexpectedly opened to the entire world simply because developers forgot to enforce authorization.

5. **Thread-safety**: As we saw earlier with the singleton example, we can quickly end up in a situation where thread safety is not guaranteed. It is important to track this down early in the life cycle of the application because it can be hell to troubleshoot later in production.

6. **Object lifetime**: It is important to check that objects are scoped correctly. New instantiations of objects that could be shared undoubtedly lead to increased CPU and memory usage. Conversely, objects that should not be shared should of course not be used as singletons (or static objects).

7. **SOLID compliance**: This means checking if the code is following the SOLID principles, among which is making sure that dependencies are handled properly. Whether you do clean architecture or not, dependencies should always be handled gracefully.

8. **Proper exception handling**: Although none of the code samples proposed in this book have proper exception handling, this is key to manage exceptions correctly. In the same vein, it is important to follow conventions such as returning an adequate HTTP code when developing an API.

9. **Scalability**: There are multiple ways to handle scalability but one of them is certainly using mechanisms such as pub/sub and leveraging asynchronous patterns whenever possible. This is mostly applicable to business assets where high load/volume is anticipated.

10. **In-memory caching, session state, and more**: I try to avoid these like the plague because they do not fit well with a scale-out story. We will see in our next chapter how cloud and cloud-native apps are a game-changer from that perspective.

As you can see, I focus more on the performance/security/reliability aspects than on readability and extensibility. The reason why I focus on this first is that the visible observations of such issues do often appear in production or when the system is under high load, which makes it trickier to troubleshoot. This top 10 list is not written in stone and may vary according to the solution.

Summary

It is a no-brainer that design patterns must be mastered by software architects because they are part of their toolbox. However, I also tried to stress the fact that you should always exercise your own judgment in your own specific context. I advocate for a pragmatic software architecture practice. Remember that, contrary to some of the clean architecture rantings, the code is not the only place to look. A good software architect aims to achieve both *fit for purpose* (where clean architecture can certainly play a role), and *fit for use*, which is most of the time ensured by non-code layers.

In our next chapter, we will see that most modern applications are often heavily distributed and rely on existing components. This is what I call the growing importance of the ecosystem.

Section 4: Impact of the Cloud on Software Architecture Practices

Getting older is never fun, but I started my career long before the cloud era and I have transitioned to a cloud architect in the meantime. What I have learned for sure is that the cloud is a game-changer in software architecture because it is often the enabler of a larger digital transformation. The cloud revamped how infrastructures get provisioned and how to address non-functional requirements in general. In this section, I propose that you review some popular architectural patterns through the prism of AWS and Azure, the biggest public cloud providers in 2021.

This section comprises the following chapter:

- *Chapter 6, Impact of the Cloud on the Software Architecture Practices*

6

Impact of the Cloud on the Software Architecture Practice

In this chapter, we will focus on one of the most recurrent topilcs of the last decade – the cloud. The cloud is a game-changer in the IT landscape and is an enabler of a larger digital transformation, which often hits organizations like a storm. Beyond technology, the cloud and cloud-native approaches require organizational changes and a proper culture to leverage all their benefits. But this goes far beyond the scope of this chapter, in which we will limit ourselves to evaluating the cloud's impact on software architecture.

More specifically, we will cover the following topics:

- Introducing cloud service models, the cloud, and cloud-native systems
- Mapping cloud services to architecture styles and patterns
- Reviewing cloud-native patterns

By the end of this chapter, you should have a better understanding of cloud and cloud-native approaches, which gain in popularity year after year. It has become essential for a software architect to jump on the bandwagon and grasp the importance of the ever-growing ecosystem.

Technical requirements

If you want to practice implementing the explanations provided in this chapter, you will need **Visual Studio 2019** to open the solution provided on GitHub.

All the code samples and diagrams for this chapter are available at `https://github.com/PacktPublishing/Software-Architecture-for-Humans/tree/master/CHAPTER%206/`.

Introducing cloud service models, the cloud, and cloud-native systems

The prevalence of cloud ecosystems is a game-changer, and you should take this into account from the start when you design applications, should you envision the cloud as a hosting platform. As per my real-world observations, this fact is often misunderstood or overlooked by software architects, who tend to neglect the ecosystem their code is running in and focus too much on the code itself. The reason why it is important to grasp the ecosystem is that it comes with pre-built services and functionalities, which can boost your productivity. The type of service model you work with tends to steer your design choices. The following diagram shows the most important service models:

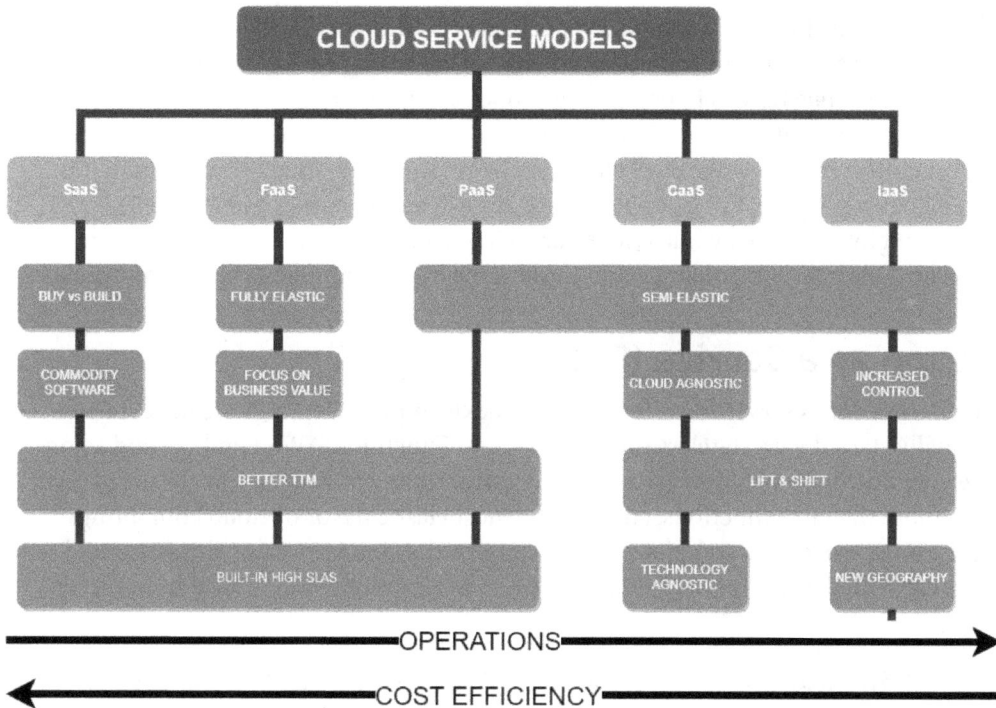

Figure 6.1 – Cloud service models

In the preceding diagram, from left to right, the level of operations increases. For example, the **Software as a Service (SaaS)** model is a fully managed off-the-shelf software offering, with embedded functional features, and comes with very low operating costs. On the opposite side, you can find **Infrastructure as a Service (IaaS)**, which is what I call *business as usual*, in the cloud, where the level of operations is *almost* equivalent to on-premises systems. The same applies to cost efficiency (TCO) but in the opposite direction, from right to left, where IaaS is not especially cost-friendly, while SaaS is the most cost-effective way to fulfill transversal (commodity) enterprise needs. Now, let's review the different models from a technical perspective.

Software as a Service (SaaS)

SaaS platforms come with many APIs that facilitate integration scenarios. **Office 365**, **Salesforce**, and so on all ship with rich APIs. SaaS often answers the *buy versus build* question, where it is almost always better to buy SaaS than trying to build something equivalent in-house. The level of operations is almost zero, except for the SOC, which actively monitors the usage of SaaS platforms and works on preventing data leakage.

Many organizations invest in a **Cloud Access Security Broker** (**CASB**) to struggle against SaaS proliferation and control user behavior regarding SaaS platforms. With SaaS systems, the challenge resides in organizing proper change management and making sure employees and collaborators do not disclose company information unexpectedly. A software architect may leverage SaaS platforms by reusing them as application building blocks whenever possible. For example, Office 365's Teams channels can be used as an entry point by end users, while Office 365's Graph API can be leveraged by custom components. The goal is to integrate and reuse some SaaS components in a larger solution.

Function as a Service (FaaS)

FaaS is also known as serverless, which we will tackle in more depth in the next chapter. FaaS initially started with stateless functions (Azure Functions, AWS Lambda, and so on) that were executed on shared multi-tenant infrastructures. Nowadays, FaaS has expanded to much more than just functions, and it is the most elastic flavor of cloud computing. While the infrastructure is completely outsourced to the cloud provider, the associated costs are calculated based on the actual resource consumption. FaaS is ideal in numerous scenarios:

- **Event-driven architectures**: Subscribe to event publishers and trigger activities accordingly. For example, having a function be triggered by the arrival of a message on a message broker, parsing it, and notifying other processes about how that message should be handled if needed.

- **Batch jobs**: You might trigger one-shot containers in a serverless way, to handle recurring activities. Once the containers have been completed, they can be fully destroyed by the underlying serverless service.

- **Asynchronous scenarios of all kinds**: FaaS is particularly suited for asynchronous activities because they do not require very low latency. Indeed, one downside of FaaS is that the cloud provider must dynamically allocate the necessary compute resources when needed, which may cause a short initialization delay. Higher latency is usually well tolerated in asynchronous scenarios, so we can live with this downside.

- **Unpredictable system resource growth**: When you do not know in advance what the usage of your application is, but you do not want to invest too much in the underlying infrastructure, FaaS helps absorb this sudden resource growth in a costly fashion. This is what I mean when I say that FaaS is fully elastic: you scale from 0 to what is necessary, then back to 0 again.

FaaS allows cloud consumers to focus on building their applications without having to worry about system capacity. Therefore, developers and software architects can focus more on the actual business value they produce. The price to pay for the flexibility and elasticity of FaaS is its disruptiveness toward traditional IT practices. Because FaaS is fully dynamic, you have low, and sometimes no, control over the network perimeter (impacting the security NFRs), which is abstracted away by the cloud provider. FaaS is therefore a source of tension for traditional IT practitioners, who are still legion.

Platform as a Service (PaaS)

PaaS is a fully managed service model and has much broader coverage than FaaS, which helps you build new solutions (or refactor existing ones) much faster. PaaS reuses off-the-shelf services that already come with built-in functionalities and whose underlying infrastructure is fully outsourced to the cloud provider. PaaS is also quite disruptive toward traditional IT, but less than FaaS. The reason why PaaS is a little less disruptive is that you pre-pay for the compute, giving you more control over it. It can also often be dedicated (non-multi-tenant), which gives you even more control. PaaS gives you more configuration options than FaaS. PaaS is semi-elastic (not fully) because you pay for pre-allocated compute. Auto-scaling plans can be defined but often remain the duty of the cloud consumer, where you do not even have to worry in the FaaS world.

Multi-tenant offerings remain more cost-friendly than dedicated ones because you can leverage economies of scale.

PaaS is suitable for many scenarios, including the following:

- **Greenfield projects**: Because PaaS is also disruptive, it is always a good idea to start with a brand-new project.

- **Internet-facing workloads**: Public cloud providers are the ideal partners when it comes to building and hosting internet-facing assets. They compare favorably with the old-school but never-dying on-premises **Demilitarized Zone (DMZ)**, which is the traditional way of exposing workloads to the internet.

- **Modernization of existing workloads**.

- **API-driven architectures**: We will look at these in the next chapter, but PaaS providers all ship with API management solutions, which greatly facilitate the development and deployment of APIs. An excellent use case for API-driven architectures is when you build your own SaaS platform to sell to other companies. We will tackle API architectures in the next chapter.

- **Anytime, anywhere, and any device scenarios**: Cloud providers usually offer high SLAs and many different flavors, which accommodate any device at any time from anywhere, out of the box.

PaaS also boosts productivity and time-to-market because developers and architects can reuse existing building blocks, which they can assemble and incorporate into their applications.

Containers as a Service (CaaS)

Containerization has become mainstream, and cloud providers could not miss that train. CaaS consists of making orchestrator platforms such as K8s and Red Hat OpenShift available to cloud consumers. In addition, cloud providers ship with proprietary offerings such as **Azure Container Instances (ACI)** and **AWS Elastic Container Service (ECS)**.

CaaS is suitable for the following scenarios:

- **Lift-and-shift**: While transitioning to the cloud, a company might want to simply lift and shift its assets, which means migrating them as containers. Most assets can be packaged as containers, without us having to refactor them.

- **Cloud-native workloads**: You can leverage the latest cutting-edge and top-notch K8s features and add-ons. We will look at cloud native in the next section.

- **Batch, asynchronous, or compute-intensive tasks**: For example, ACI and ECS can both accommodate batch jobs.

- **Portability**: CaaS offers greater portability than anything else, and it helps reduce the vendor lock-in risk to some extent.

- **Service meshes**: Most microservice architectures rely on service meshes, which, in turn, rely on containerization platforms, in their modern form.

- **Modern deployment**: CaaS offers modern deployment techniques, such as A/B testing, canary releases, and blue-green deployment. These techniques prevent and reduce downtime in general.

CaaS is closer to the infrastructure and is often less managed than FaaS and CaaS. The cloud consumer does more operational work, such as upgrading cluster versions, patching nodes, and so on. This varies from one cloud provider to another.

Infrastructure as a Service (IaaS)

IaaS is the least disruptive service model. It is the process of renting a data center to a cloud provider. It is business as usual in the cloud. IaaS is not the service model of choice to accomplish a digital transformation, but nevertheless, it can be useful in numerous scenarios:

- **Lift-and-shift**: Because on-premises systems also use virtualization everywhere, it is very easy to deploy existing applications to the cloud, while not modifying anything in the applications themselves.

- **High-performance computing** (**HPC**): Most cloud providers have impressive HPC capabilities that can be made available immediately and at a much higher scale than what most companies can afford on-premises.

- **Small companies with high compliance requirements**: IaaS can be a good alternative for smaller companies that do not want to invest in their own data centers, and if they have high compliance requirements that could not be fulfilled by other service models.

- **Disaster recovery**: IaaS is very suitable for disaster recovery purposes because cloud providers come with tooling that facilitates such scenarios.

- **Compute shortage or end-of-life hardware**: When you are short on compute in your own data center(s) or are confronted with end-of-life hardware, it may be easier and faster to take the IaaS path.

- **New geography**: If your business spans new geography, it could be faster to start it in the cloud while inheriting from the cloud provider's compliance with local regulations.

With regard to costs and operations, IaaS is almost equivalent to on-premises, although it is very hard to compare the TCO of IaaS and on-premises.

Of course, facilities, physical access to the data center, and more are all managed by the cloud provider. It is not necessary to buy and manage the hardware yourself anymore.

Anything as a Service (XaaS or *aaS)

Other service models exist, such as **Identity as a Service** (**IDaaS**) and **Database as a Service** (**DBaaS**), to such an extent that the acronym **XaaS**, or ***aaS**, was born around 2016, to designate all the possible service models. It is important for a software architect to grasp these different models as they serve different purposes, require different skills, and directly impact the application you build. These service models are also likely to impact some software quality attributes, as we will see in the next section.

Service models and software quality attributes

Choosing a service model is not neutral. As we saw in the preceding sections, it has an impact on costs, operations, required skills, and enabling (or not) some technical patterns. It also impacts software quality attributes. The following table shows how FaaS, PaaS, and CaaS impact some recurring attributes:

	FaaS	PaaS	CaaS
Scalability	+++	++	+++
Availability	+++	++	++
Deployability	+	+	+++
Portability	-	-	+++
Testability	+	+	++
Security (control)	-	+	++
Auditability	+	+	+
Reliability	++	++	++
Manageability	+++	+++	+

Figure 6.2 – Impact of service models on quality attributes

We probably don't need to mention that a + sign represents a positive impact and a – sign represents a negative one. Admittedly, this list of attributes is not exhaustive and the scores are subject to debate, but this is, in any case, not neutral. You may refine or augment this list for your own context. You must balance this with the level of operations and costs that pertain to each service model. Now that we have reviewed the most important service models, let's look at what cloud and cloud native mean.

Cloud versus cloud native

Let's start by introducing the notion of cloud and cloud-native development, which you might not be familiar with. The cloud is a particular ecosystem that has unique capabilities. If we want to leverage these capabilities, we must rethink the way we design applications and the types of frameworks we should consider.

Let me directly evacuate IaaS from the equation because as we stated previously, IaaS is business as usual in the cloud. It is a replication of your on-premises systems in the cloud. Many on-premises applications are restricted to an N-tier architecture, with a frontend, backend, and database, living on a few servers. We control the entire code base and underlying systems *because we do everything ourselves*. We write custom in-house frameworks to deal with cross-cutting concerns, such as logging, exception handling, and so on. Our in-house frameworks become monoliths after a while. From that perspective, IaaS is fine but is by no means cloud or cloud native. You can design an application for IaaS the same way you design applications on-premises. IaaS does not come with any specific constraints or pre-built functionality, which is why it has no impact on software architecture. Cloud and cloud-native approaches imply a mindset shift, which is not required nor even facilitated by IaaS.

Clarification made, one of the biggest challenges when starting to work with cloud and cloud-native approaches is that applications are *distributed and rely on existing services*. In traditional IT, we still have many monoliths, or in the best case, we work with **service-oriented architecture (SOA)**, which we talked about in *Chapter 4, Reviewing the Historical Architecture Styles*. SOA has proven its value and is certainly still future-proof, but as always, the IT world has evolved, and newer paradigms have emerged.

With cloud and cloud-native applications, we tend to rely more on off-the-shelf frameworks, services, and ecosystems. We *assemble* existing cloud services and add our code on top. Some services come with a lot of built-in functionality. For example, many mainstream artificial intelligence services exist that can be used in applications to perform image recognition, text extraction, and so on. Other services also come with more technical features. For example, **API Management** solutions ship with API throttling, JWT token validation, and many other features out of the box. Moreover, such services are designed with resilience and robustness from the ground up. You should, at best, create a pale copy, should you stubbornly develop it yourself.

To adopt a cloud or cloud-native approach, you need to modernize/refactor the assets or start from a greenfield situation. As a software architect, you must look at how the ecosystem can help you achieve more faster and in a better way. As a software architect, you must offload the *reusability* quality attribute to the provider. You must pay particular attention so that you don't reinvent the wheel.

Although there is probably not a single definition for cloud and cloud-native development, let me share my definitions. A **cloud development approach** looks as follows:

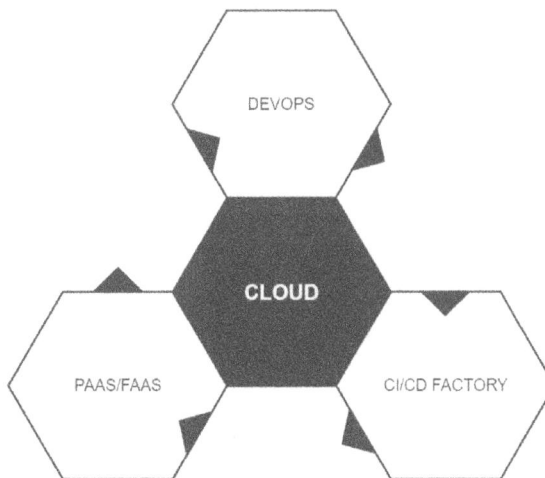

Figure 6.3 – Cloud development approach

Cloud development relies on three pillars:

- **DevOps**: This is the way we organize teams to collaborate on a product. Beyond technicalities, it touches organizational aspects, such as the notion of virtual teams, the product owner, **minimum viable product (MVP)**, user stories, sprints, and so on. The purpose of DevOps is to enhance collaboration and align the different stakeholders. There is a famous saying that reflects this mindset: *you build it, you run it!*

- **CI/CD**: The automation toolchain plays a crucial role in setting things to music. The **Infrastructure as Code (IaC)** approach is an integral part of cloud and cloud-native development. Software architects must grasp the capabilities of IaC and make sure infrastructure architects grasp it too.

- **PaaS/FaaS**: I explained these hosting models in the previous section. They allow us to leverage IaC to boost our productivity. With a proper automation toolchain, we can easily create new environments or delete them. We do not need to rely on the traditional **Development Test Acceptance Production (DTAP)** anymore. We can spin up and delete environments as needed. Mindset-wise, you must accept that you cannot control everything. As a software architect, you analyze the ecosystem and try to leverage it as much as possible. You can, of course, try to add some abstraction layers to your code to prevent vendor locking, but you should not abuse them. When it comes to cloud and cloud native, *leveraging is better than abstracting away*.

A **cloud-native development approach** looks as follows:

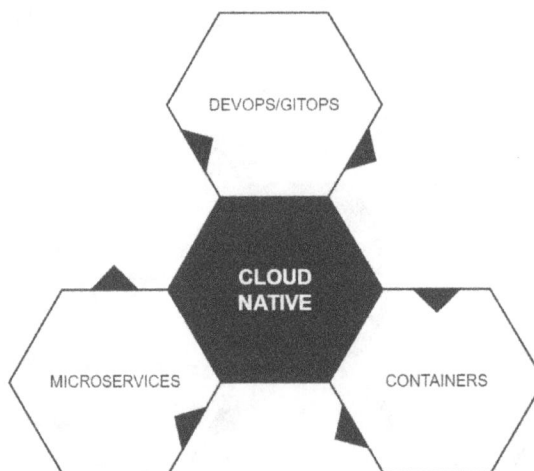

Figure 6.4 – Cloud-native development

At first glance, it looks very similar because *it is* very similar:

- **DevOps/GitOps**: The cultural aspects remain the same, but the tooling differs. Most cloud-native factories rely on GitOps instead of DevOps. GitOps revamps the way infrastructures and applications are deployed. It targets container platforms such as Kubernetes.

- **CaaS/Containers**: They represent the main difference between cloud and cloud-native approaches. Strangely, you can work in a cloud-native way on-premises, should you have your own K8s or OpenShift clusters, and if the clusters themselves can be fully deployed and configured automatically.

- **Microservices**: In *Chapter 4, Reviewing the Historical Architectural Styles – Monoliths, SOA, and Microservices*, we explained that microservices are often based on container platforms because they offer the best possible support. *This does not mean that non-microservice applications cannot be cloud native*, but microservices leveraging container platforms are cloud native by design.

Containers and tooling are the biggest differences between cloud and cloud-native approaches. However, they share the same DNA. They both aim at deploying faster, gaining agility, and gaining autonomy while optimizing costs. They both require a mindset shift, especially if you run them at scale. They are both very disruptive toward traditional IT practices and software architecture. They both force the software architect to consider the ecosystem as an enabler to achieve faster and better results. Finally, they both have a positive impact on many quality attributes. Now, let's map some cloud services to architecture styles and patterns.

Mapping cloud services to architecture styles and patterns

To make sure you realize the importance of the ecosystem, I have mapped a few typical cloud architecture styles and design patterns to some cloud services. These services help you achieve results faster and better. You can rely on them to boost your productivity and comply with many NFRs out of the box. This does not mean that you cannot work with similar patterns on-premises, nor that you are restricted to the services depicted in *Figures 6.5 and 6.6*, but this should give you a solid overview.

To illustrate the mapping between the patterns and the services, I considered **Microsoft Azure** and **Amazon Web Services (AWS)** because they are the two leading cloud providers at the time of writing:

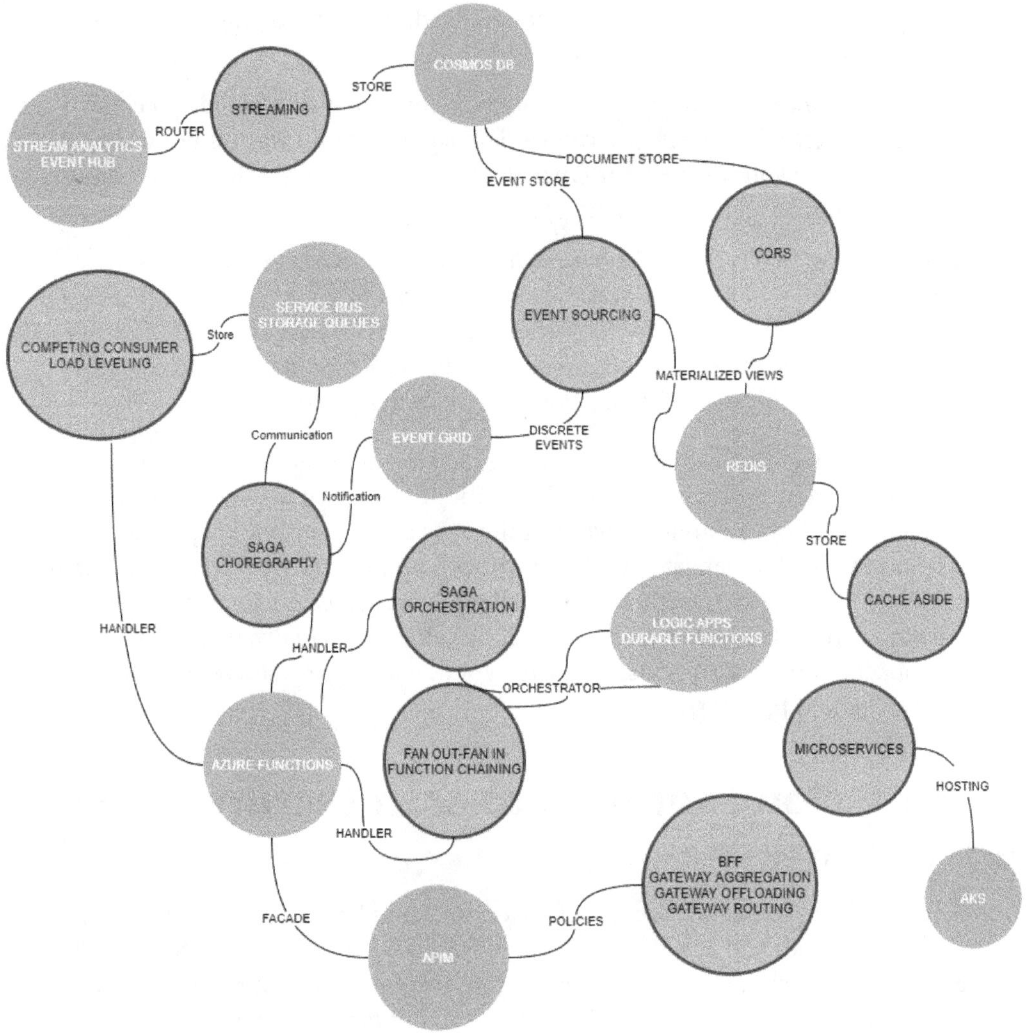

Figure 6.5 – Azure services mapped to patterns

The circles with a thick border are the patterns, while the other circles represent the cloud services. For example, if you must build a SAGA (which I will explain in the next section) in choreography mode, you can rely on Azure Service Bus and/or Azure Event Grid to handle the communication between the different SAGA participants. If you understood this, you should be able to read the preceding diagram easily. I did the same exercise with AWS, as illustrated in the following diagram:

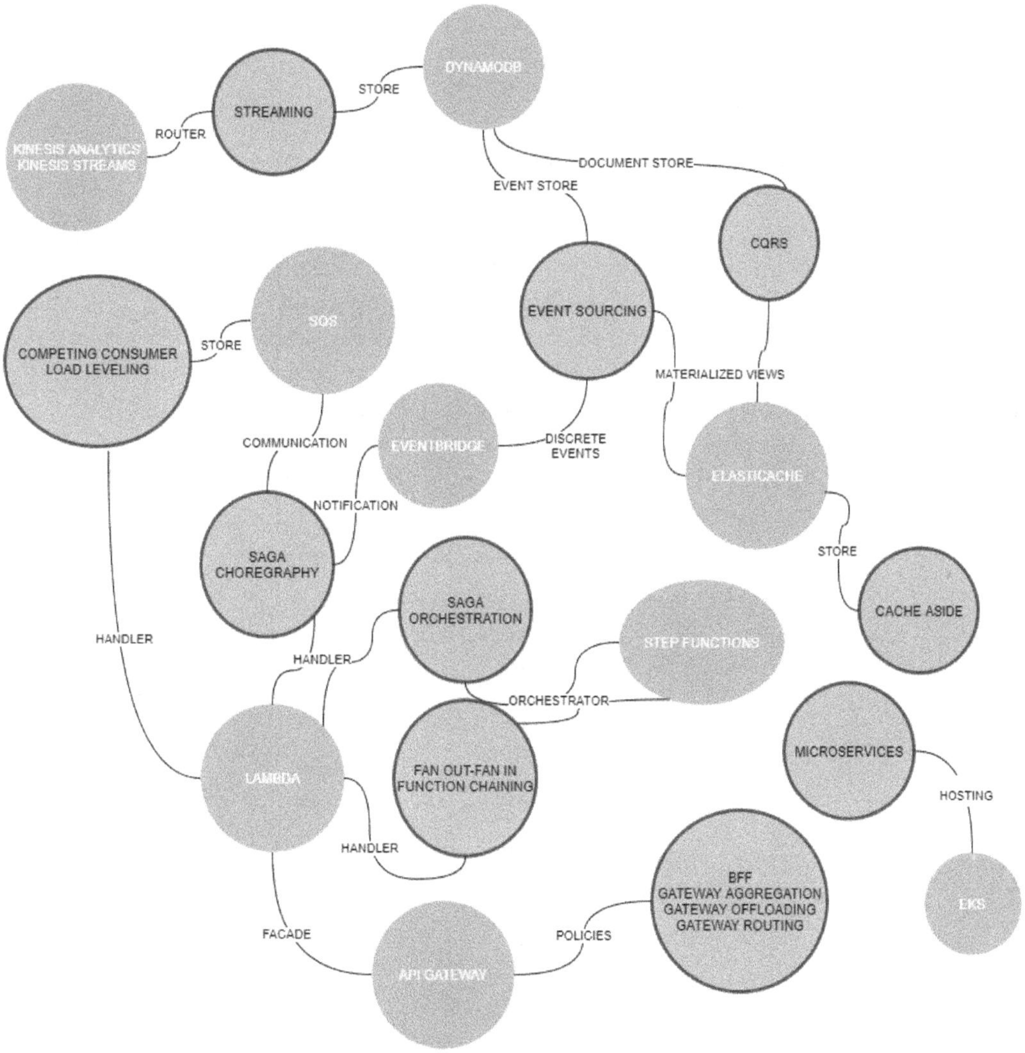

Figure 6.6 – AWS services mapped to patterns

The preceding diagram is built on the same principles as *Figure 6.5*. As you can see, cloud services facilitate many design patterns. Now, let's review some of the patterns depicted in the two preceding diagrams in more detail.

Reviewing cloud and cloud-native patterns

Now that you have more clarity on the service models and their high-level impact, let's explore some patterns in more depth.

The Cache-Aside pattern

With cloud and cloud-native patterns, scaling a component is a scale-out/in story, not a scale-up/down one. This means that we multiply the number of instances, instead of adding more compute to a single instance. Scaling up/down is, of course, still possible, but scaling out/in is by design. Multiplying instances is not a neutral thing because it may disturb every in-process thing. Therefore, it has an impact on how we can handle data caching, user sessions, and so on. When considering a scale-out story, you should try to avoid in-process caching/sessions. Therefore, the Cache-Aside pattern should be implemented with an external cache system, such as **Azure Redis Cache** or **AWS ElastiCache**. The goal is to share cached data across instances and to prevent inconsistencies. The following is a representation of the Cache-Aside pattern:

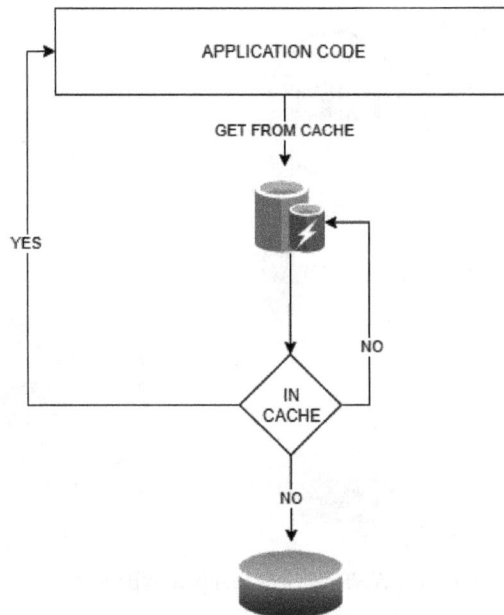

Figure 6.7 – Cache-Aside pattern

The code checks whether a given value is in the cache or not. If it's not in the cache, it goes to the data store and updates the cache for future use. If the item is already in the cache, it does not reach out to the data store. The Cache-Aside pattern also improves performance. In the preceding diagram, we have a data store, but you can also use the cache store alone. Now, let's look at another frequent pattern in the cloud, namely the SAGA pattern.

Understanding the SAGA pattern

The **SAGA** pattern deals with distributed transactions. This means that there are multiple participants involved when considering a transaction as completed. In the **Atomicity Consistency Isolation Durability (ACID)** world, a transaction is an atomic group of operations. They all succeed, or they all fail, at once. This works well with monoliths when a single backend writes to a SQL database, but this is not applicable to distributed applications, which involve more components to consider a transaction as complete. With microservices, the segmentation and segregation of duties across services themselves cause the transactions to be distributed by design. Also, most of the time, each microservice has its own data store, which can even be based on different engines. In such an architecture, it is impossible to rely on ACID anymore.

Additionally, cloud and cloud-native implementations often rely on FaaS and PaaS services, which sometimes do not even encompass the transaction concept. So, if you cannot use database-level transactions, you must rely on a different mechanism, such as the SAGA pattern. Unlike ACID, SAGA cannot simply roll back the whole transaction because of one failed operation; instead, it brings the concept of compensating transactions. Local transactions that have been committed by local services are already persisted to their own data store. If one of the participants fails, the mechanism of compensating transactions should make them invalidate what was committed, or resume to a certain sequence item.

SAGA is an orchestrator-based or choreography-based pattern. The following diagram represents the orchestrator pattern:

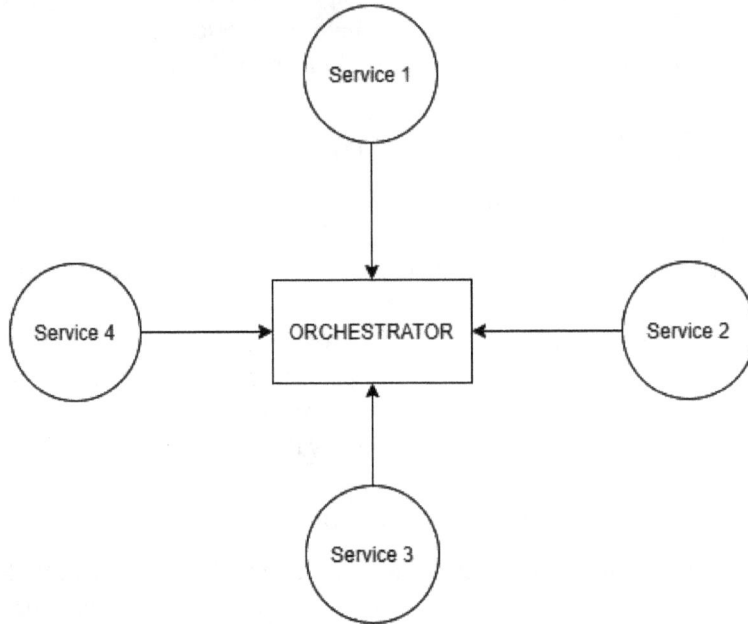

Figure 6.8 – SAGA orchestrator-based pattern

The orchestrator is in the driving seat and orchestrates the different participants that are part of a transaction. Each service is unaware of what other services do, and only the orchestrator knows the actual state of the ongoing transaction. If one of the local steps fails, the orchestrator will trigger one or more compensating transactions to invalidate the preceding steps when required.

In a choreography-based pattern, as shown in the following diagram, services are bridged together with a pub/sub mechanism, where each service publishes its outcomes (success or failure) that one or more subscribers capture, and, in turn, publish events about their own outcome:

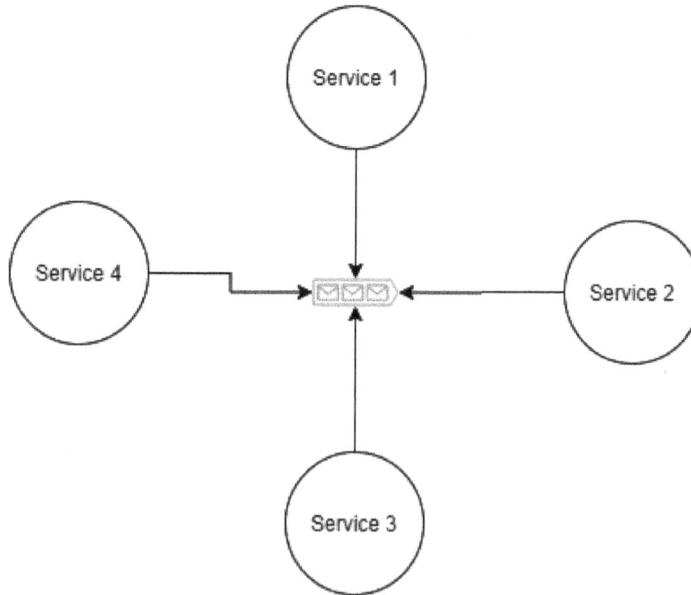

Figure 6.9 – SAGA choreography-based pattern

If one of the participants fails, it will publish its failure event, which will be captured by the others to invalidate their local transaction. However, the choreography pattern makes it harder to understand the full picture when a single transaction involves many participants. You can see an example of SAGA using Azure services at `https://github.com/Azure-Samples/saga-orchestration-serverless`. Now, let's look at CQRS.

Command Query Request Segregation (CQRS)

In a nutshell, **CQRS** is a pattern that segregates commands (something that mutates state) and queries (something that returns data without mutating state). CQRS is often confused with **Command Query Segregation** (**CQS**). The latter is the strict definition of the preceding definition. This is because you can segregate queries and commands even when targeting the same data store. True CQRS implies *request segregation*, which means that very different systems and data stores can handle requests and commands. This gives you the ability to increase the speed of both write and read operations and be able to scale them independently. By splitting the data stores (or having read-only and write-only partitions), heavy read operations do not impact the write performance and vice versa.

As you might have guessed already, working with different data stores automatically implies eventual consistency, hence true CQRS is not suitable for every application because not every application supports eventual consistency. I see many developers and software architects replace the plain **Create Read Update Delete** (**CRUD**) pattern with CQS (although they think they're using CQRS), but this brings lower value and is often overkill since mere CRUD APIs do the job perfectly fine most of the time. Yet, both the CQS and CQRS approaches force you to think about your command and query needs upfront, which then helps you engineer your database(s) appropriately. This statement is particularly true for NoSQL stores, which are often more challenging to engineer properly. The following diagram shows a simplified CQRS implementation:

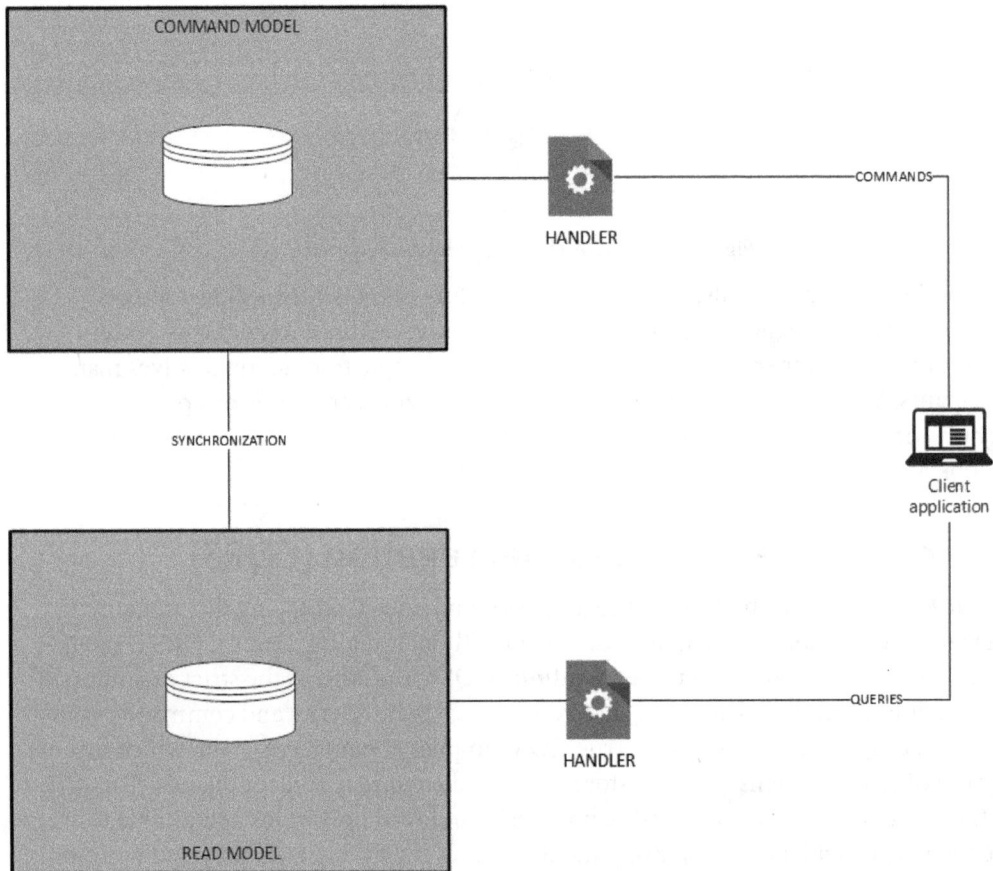

Figure 6.10 – The CQRS pattern

Here, you have a command and a read model, both relying on their own data stores. The client application talks to different channels when querying data or mutating state. A synchronization process happens between the command model and the read one and varies according to the needs. In very basic scenarios where you simply want to split reads and writes, you may use database read-only replicas and let the database engine synchronize the data for you. In more advanced scenarios, the read model is composed of **materialized views**, which are pre-built query results that are computed by the synchronization process. Queries are very fast because results are tailor-made for query-specific requirements. When applied this way, the CQRS pattern shows its value. As I mentioned previously, cloud apps are distributed by nature, and that is why CQRS is more commonly used in the cloud than on-premises.

A single application may ship with read-only APIs and command-only APIs. You can regroup both requests and commands under the same API, but this will require splitting commands and queries through a CQS approach first, and then letting each mode talk to its own data store, which converts CQS into CQRS. In *Chapter 5, Design Patterns and Clean Architecture*, we learned how to use a repository together with an ORM. When applying CQRS to a single API, you may combine the **mediator** and **repository** patterns. The following diagram shows the different components assembled:

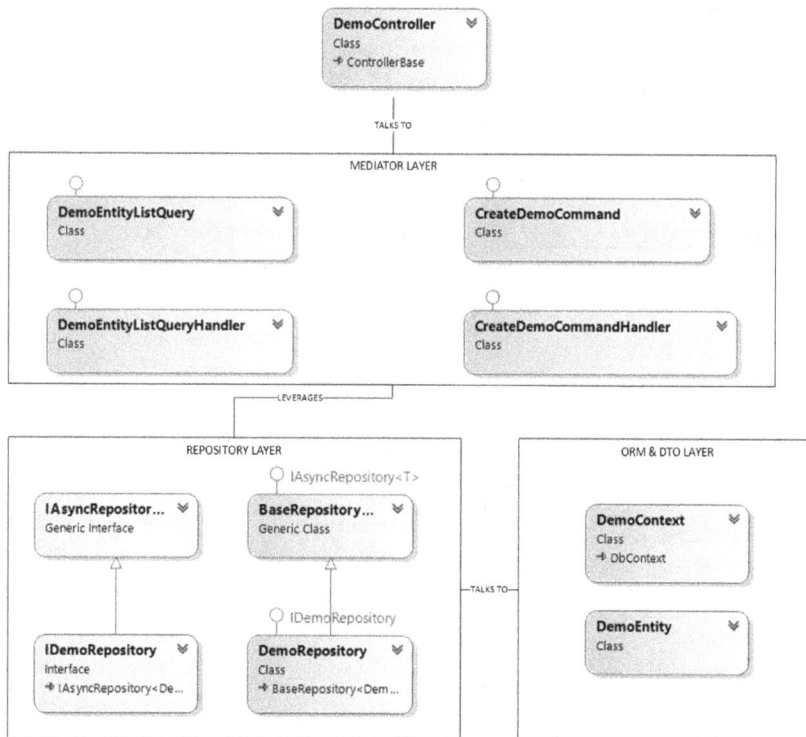

Figure 6.11 – Mediator combined with repositories to achieve CQS or CQRS served by a single API

If your query-related repositories talk to a read model and your command-related repositories to a write model, then you have achieved CQRS. Of course, your read model should also be updated whenever data is mutated. Refer to the service mappings (*Figures 6.5* and *6.6*) depicted in the previous section to figure out which services can help you build the CQRS pattern. You can also visit the GitHub repository mentioned in the *Technical requirements* section to see a fully working CQS implementation.

Event sourcing

Event sourcing is always used in conjunction with CQRS because every implementation makes use of materialized views for the read model. So, again, writes and reads are separated, but the pursued objective differs, since the primary purpose of event sourcing is to provide an audit trail and extreme scalability. Since events are immutable, we can understand the full history of the data at any point in time. Even better, we can rebuild the actual data state by replaying events from the event store. The following is a simple event sourcing diagram:

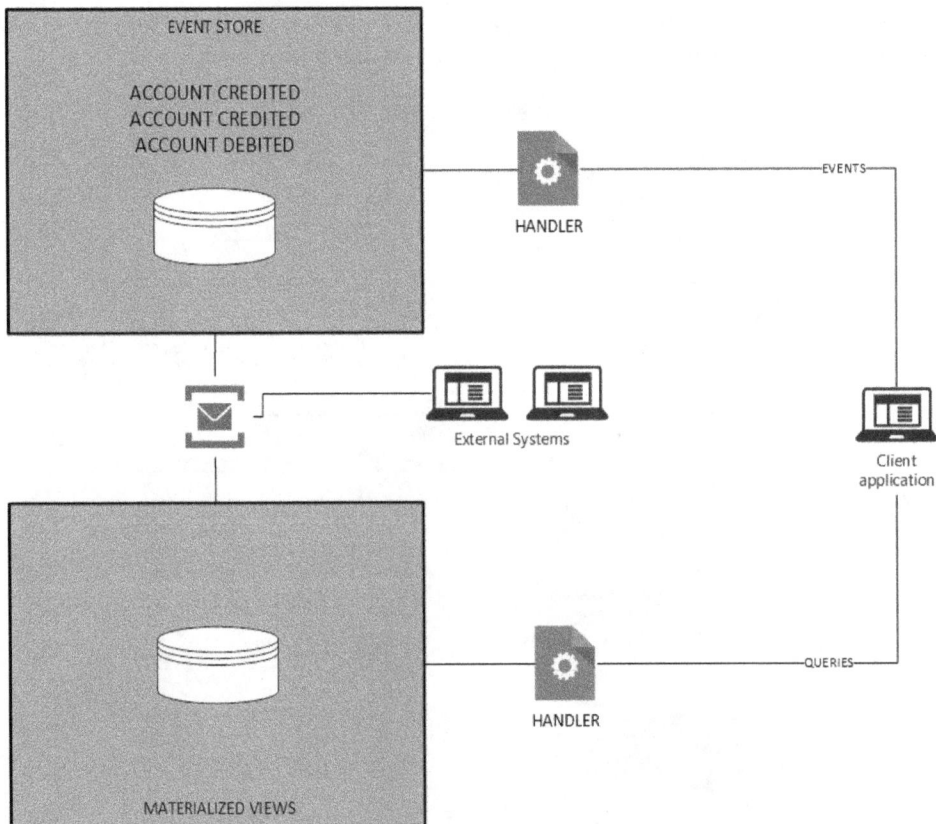

Figure 6.12 – Event sourcing diagram

The event store holds the entire history of the events that occurred during the life of an application. Once validated, every event is published to a message broker (or event bus) to build materialized views. External systems can also subscribe to the published events to handle them in their own context. With event sourcing, the event store is the single source of truth, which can be used for audit purposes as well as to rebuild a new instance of the app at any point in time by replaying the events. Event sourcing is an *extremely complex* pattern that should only be used when it adds value to your business case.

Summary

The message I wanted to convey in this chapter is that cloud and cloud-native applications are very disruptive with regard to traditional IT practices and design patterns. Cloud service models and container platforms, such as Kubernetes, are broad ecosystems that you should leverage in your application and solution designs. As we saw in the *Service models and software quality attributes* section, the cloud service models have a positive impact on many attributes, by design. We also saw that concrete pattern implementations can be achieved faster and better, thanks to cloud services. Finally, we delved into some recurrent cloud and cloud-native patterns that are inherent to distributed applications.

In the next chapter, we will focus on API-driven architectures, both serverless and microservices-based, which are often built using the cloud and cloud-native systems.

Section 5: Architectural Trends and Summary

Theory is good, but sometimes a little bit of practice helps digest the theoretical concepts. In this last section, I want to give you a more concrete technical experience of trendy architectures by going through a microservice and a serverless use case. Although I had to choose some technologies to build a concrete example, be sure that the principles explained throughout the book go beyond that choice.

This section comprises the following chapter:

- *Chapter 7, Architectural Trends and Global Summary*

7
Trendy Architectures and Global Summary

This chapter is the continuation of the previous one since trendy architectures are often cloud-based. In this chapter, we are going to focus on the main architectural trends that every software architect should master.

More specifically, we will cover the following topics:

- API-driven architectures
- Hands-on with a microservice architecture example
- Hands-on with a serverless architecture example

Because I have already introduced the theoretical aspects of serverless and microservices, I wanted to satisfy the developer in you, and walk you through more concrete examples, at the risk of being less technology agnostic. By the end of this chapter, you should have grasped the basics about API-driven architectures and get started with serverless and microservices.

Technical requirements

If you want to practice implementing the explanations provided in this chapter, you will need the following:

- **Visual Studio 2019**: To open the solution provided on GitHub.

- **Kubernetes**: You will need a vanilla cluster such as **MiniKube** or **Docker Desktop** with K8s embedded. You can also use any cloud-provided cluster (Azure, AWS, or GCP). I used Azure Kubernetes Service to host my demo solution. Whatever solution you choose, make sure that the cluster has access to the internet so that it can pull the Docker images that I published to **Docker Hub**.

- **An Azure subscription**: I used Azure for the serverless sample application. To create your own free Azure account, follow the steps explained at `https://azure.microsoft.com/free/`.

All the code samples and diagrams for this chapter are available at `https://github.com/PacktPublishing/Software-Architecture-for-Humans/tree/master/CHAPTER%207/`.

API-driven architectures

Modern assets ship with APIs, but the notion of an API itself has changed over time. In the nineties, an API was some sort of client library you could use to interact with an application. In 2021, an API is a physical endpoint acting as a client interface that allows clients to interact with backend services. The form has changed but the purpose is the same. Both forms aim at facilitating integration scenarios and exposing application features to client programs. Yet, I have noticed that although developers are usually aware of how to develop backend services, they often lack skills in API management solutions that can be used for both internal and internet-facing APIs. API management solutions accommodate a few design patterns out of the box:

- **Backend for frontends** (**BFF**): The purpose of a BFF is to propose an API that is tailor-made to a given consuming channel, such as a mobile application. The purpose is to satisfy the specific requirements of a given client. A mobile app is often (less than before) limited in bandwidth and has smaller screens, so you may not want to return the same amount of information than what you would for a regular website. Similarly, you may want to filter out a few API operations, depending on the client consumer. API management solutions also offer response transformation mechanisms.

- **Gateway aggregation**: This pattern is quite close to the *facade pattern* but instead of doing things in code, you rely on API gateway policies to do the heavy lifting. Note that this pattern can also be easily tackled with code-based technologies such as **GraphQL** (`https://graphql.org/`).

- **Gateway offloading**: The purpose of this pattern is to offload cross-cutting concerns such as mutual authentication, JWT token validation, TLS termination, throttling, data caching, and so on to an API gateway. This prevents you from writing code for such activities and making sure requests that reach your backend services have gone through a series of verifications first. Every illegal request gets discarded by the gateway. This not only makes your life easier as a developer, but it also makes your application more robust and secure. Cloud-based API gateways have **Denial of Service (DOS)/Distributed Denial of Service (DDOS)** mitigations built in.

- **Gateway routing**: API gateways make it possible to route incoming requests dynamically to different backend services according to the input request and context variables. Most cloud-based gateways can route a request to the closest region (closest to the client app) and/or fastest backend service.

API management solutions help you tackle the preceding list of patterns, but there is more to this. They effectively allow you to manage the life cycle of your APIs and expose them to your internal and external customers. There are many off-the-shelf solutions, such as Azure API Management, AWS API Gateway, and MuleSoft Anypoint API Manager, to name a few. As a software architect, you cannot miss out on these types of solutions. Microservice architectures, as we'll see in the next section, are typical API-driven architectures.

Hands-on with a microservice architecture example

I introduced microservices in *Chapter 4, Reviewing the Historical Architecture Styles*, at a high-level. Going through a full explanation is beyond the scope of this book. However, because microservices have become a serious trend, I wanted to dive a little deeper with a concrete example. The objective is to focus on the communication aspects, both synchronous and asynchronous, of the different services.

The following diagram illustrates a small application I built for you to taste the flavor of microservices:

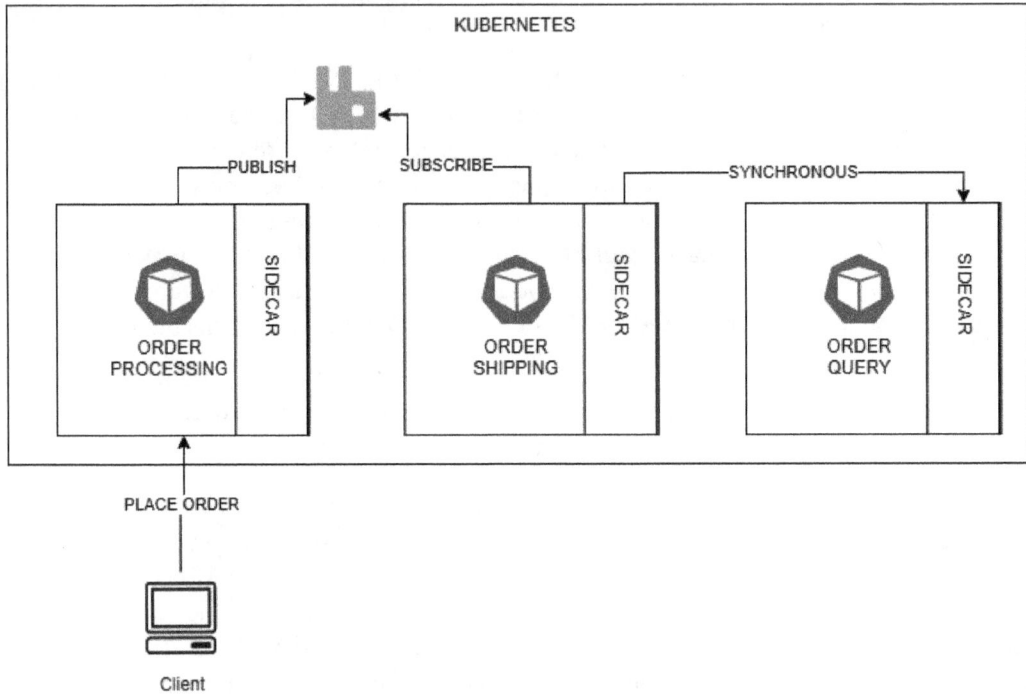

Figure 7.1 – Microservice application example

In this example, we have three services:

- The **order processing service**, which is called by the client app that places orders. This client app could be a mobile app, a web app, or another API. In theory, you would put a BFF between the client and the order processing service, but I wanted to keep things as simple as possible.

  Once the order processing service has created an order, it publishes an event to a **RabbitMQ** broker, to notify the shipping service (and potentially others) *asynchronously* about the newly created order.

- The order **shipping** service picks up every incoming order event to start the shipping process. First, it performs a *synchronous* query to the order query service to retrieve extra details about the order and starts the shipping process accordingly.

- The entire communication's plumbing (synchronous and asynchronous) is ensured by **Distributed Application Runtime** (**Dapr**), which is a technology-agnostic runtime.

All three services are hosted in a K8s cluster. You can find the full sample in this book's GitHub repository, and the required steps to run this sample app in the .README file. For your convenience, I have published the service container images to the public Docker Hub so that you do not have to worry about having your own container registry.

Now, let's focus on the essential parts of this sample application.

Service discovery and communication

As stated in the preceding paragraph, Dapr (https://dapr.io/) is used to ensure both synchronous and asynchronous service communication. I do not have a crystal ball, but I am quite sure that Dapr will become a first-class citizen in the microservices world soon. In a nutshell, it allows you to handle very common cross-cutting concerns such as pub/sub, state management, bindings, and so on.

Dapr is technology- and cloud-agnostic. You can bind dozens of cloud services (AWS, GCP, Azure, Alibaba, and so on) seamlessly, without reinventing the wheel in code yourself. Dapr illustrates the growing importance of the ecosystem, and, as a software architect, you must evaluate how such frameworks can be used as architectural building blocks. Once you have downloaded the Dapr CLI tool (https://docs.dapr.io/getting-started/install-dapr-cli/), installing Dapr is a one-liner command:

```
dapr init --runtime-version 1.0.1 -k -
```

The -k option will target the Kubernetes cluster you're currently working with.

Exploring the essential parts of the code

Now, let's explore the essential parts of the code to make sure you understand the process. The order processing service is shown in the following screenshot:

```
[ApiController]
3 references
public class OrderController : ControllerBase{
    private readonly DaprClient _dapr;
    private readonly ILogger<OrderController> _logger;

    0 references
    public OrderController(ILogger<OrderController> logger, DaprClient dapr){
        _logger = logger;
        _dapr = dapr;
    }

    [HttpPost]
    [Route("order")]
    0 references
    public async Task<IActionResult> Order([FromBody] Order order, [FromServices] DaprClient daprClient){
        //we pretend to create an order
        _logger.LogInformation($"Order with id {order.Id} created!");
        await _dapr.PublishEventAsync<Order>("bus", "order", order);
        return Ok();
    }
    0 references
    async Task<IActionResult> PublishOrderEvent(Guid OrderId, OrderEvent.EventType type){
        //we publish the order created event
        var ev = new OrderEvent
        {
            id = OrderId,
            name = "OrderEvent",
            type = type
        };
        await _dapr.PublishEventAsync<OrderEvent>("bus", "order", ev);
        return Ok();
    }
}
```

Figure 7.2 – Order processing backend service

First, we get a DaprClient instance through DI. We expose the order endpoint through an HTTP post. We pretend to create an order and then we publish the order created event through Dapr, which links to our RabbitMQ broker through a component file that I will show later.

The shipping service's implementation looks as follows:

```
[Topic("bus", "order")]
[HttpPost]
[Route("dapr")]
0 references
public async Task<IActionResult> ProcessOrderEvent([FromBody] OrderEvent ev){
    _logger.LogInformation($"Received new event");
    _logger.LogInformation("{0} {1} {2}", ev.id, ev.name, ev.type);
    switch (ev.type){
        case OrderEvent.EventType.Created:
            var order = await GetOrder(ev.id);
            if (order!=null){
                _logger.LogInformation($"Starting shipping process for order {ev.id} with " +
                    $"{order.Products.Count} " +$"products!");
            }
            else{
                _logger.LogInformation($"order {ev.id} could not be retrieved, suspending shipping process!");
            }
            break;
        other cases
    }
    return Accepted();
}
2 references
async Task<Order> GetOrder(Guid id){
    try{
        return await _dapr.InvokeMethodAsync<object, Order>(
            HttpMethod.Get,
            "orderquery",
            id.ToString(),
            null);
    }
    catch (Exception ex){//should be more specific
        _logger.LogError(ex.Message);
        return null;
    }
}
```

Figure 7.3 – Shipping service implementation

The `ProcessOrderEvent` method is decorated with a few attributes. The most important one is the first line, `Topic("bus", "order")`, which subscribes the shipping service to the order topic. Then, the `GetOrder` method is used to make a synchronous call to the order query service to retrieve extra details about the order. This method relies on Dapr's built-in service discovery to find the query service. The only thing you need to know as a service client is the application identifier, which, in this case, is `orderquery`. Of course, the code shown in the preceding screenshot is not production-ready – because you would need to handle unhappy cases – but it is enough to demonstrate the communication plumbing, which is very important in microservice architectures.

Deploying the application

To get started, you need to have your K8s cluster ready, you must pre-deploy RabbitMQ, and you must grab its default credentials (follow the steps on GitHub). Once done, you can deploy the application using the YAML deployment file shown (truncated) here:

```
apiVersion: dapr.io/v1alpha1
kind: Component
metadata:
  name: bus
  namespace: microserviceapp
spec:
  type: pubsub.rabbitmq
  version: v1
  metadata:
  - name: host
    value: "amqp://user:pwd@rabbitmq.default.svc.cluster.local:5672"
---
apiVersion: apps/v1
kind: Deployment
metadata:
  name: orderprocessing
  namespace: microserviceapp
  labels:
    app: orderprocessing
spec:
  replicas: 1
  selector:
    matchLabels:
      app: orderprocessing
  template:
    metadata:
      labels:
        app: orderprocessing
      annotations:
        dapr.io/enabled: "true"
        dapr.io/app-id: "orderprocessing"
        dapr.io/app-port: "80"
    spec:
      containers:
      - name: orderprocessing
        image: stephaneey/orderprocessing:dev
        imagePullPolicy: Always
```

Figure 7.4 – YAML deployment file

The interesting part is the first block, which deploys the Dapr component, which targets RabbitMQ through its type. The host attribute corresponds to the cluster endpoint RabbitMQ is listening to. The second block is the deployment of the order processing service (the two other services are deployed the same way). In the `annotations` section, we tell the system to enable Dapr and we give our service an application identifier.

Testing the application

Once you have deployed the app, following the prerequisites explained on GitHub, you should be able to test it. Because I wanted to keep things as simple as possible, I have not created an ingress rule, meaning that the deployed services cannot be accessed from outside K8s. Follow these steps to test the app:

1. Verify that all the pods are running:

```
Windows PowerShell
PS C:\> kubectl get pods -n microserviceapp
NAME                                     READY   STATUS    RESTARTS   AGE
orderprocessing-ff6f5644-5sjzb           2/2     Running   0          87s
orderquery-5557767d84-mcwcc              2/2     Running   0          87s
shippingprocessing-7bcfb96bfb-q6kbq      2/2     Running   0          87s
PS C:\>
```

Figure 7.5 – Listing pods

2. You should see that each pod runs two containers – the service and the Dapr sidecar. To place an order, we need to forward the order processing (or any other Dapr-injected pod) traffic to the host:

```
Windows PowerShell
PS C:\> kubectl -n microserviceapp port-forward orderprocessing-ff6f5644-5sjzb 3500:3500
Forwarding from 127.0.0.1:3500 -> 3500
Forwarding from [::1]:3500 -> 3500
```

Figure 7.6 – Forwarding traffic to the host

3. Make sure to replace the pod name with your own. Once done, we can start making calls to our localhost endpoint.

4. Using your preferred tool (Postman, Fiddler, or whichever you like), run the following HTTP POST request (a request sample is also available on GitHub):

```
Parsed  Raw    Scratchpad  Options

POST http://localhost:3500/v1.0/invoke/orderprocessing/method/order HTTP/1.1
Content-Type: application/json

{
 "Id":"4aadc0f8-eeda-4ee7-9c26-a6d39cbfbc28",
 "Products":[{"Id":"5678f982-2ae4-408c-92ff-6af45118d159"}]
}
```

Figure 7.7 – Sample HTTP POST request against the order processing service

5. Note that we could also use gRPC but, for the sake of simplicity, I simply performed an HTTP call.

6. To see whether the request was handled properly, inspect the service logs:

```
Windows PowerShell

PS C:\> kubectl -n microserviceapp logs orderprocessing-ff6f5644-5sjzb orderprocessing
[40m[32minfo[39m[22m[49m: Microsoft.Hosting.Lifetime[0]
      Now listening on: http://[::]:80
[40m[32minfo[39m[22m[49m: Microsoft.Hosting.Lifetime[0]
      Application started. Press Ctrl+C to shut down.
[40m[32minfo[39m[22m[49m: Microsoft.Hosting.Lifetime[0]
      Hosting environment: Production
[40m[32minfo[39m[22m[49m: Microsoft.Hosting.Lifetime[0]
      Content root path: /app
[40m[1m[33mwarn[39m[22m[49m: Microsoft.AspNetCore.HttpsPolicy.HttpsRedirectionMiddleware[3]
      Failed to determine the https port for redirect.
[40m[32minfo[39m[22m[49m: OrderService.Controllers.OrderController[0]
      Order with id 4aadc0f8-eeda-4ee7-9c26-a6d39cbfbc28 created!
PS C:\>
```

Figure 7.8 – Inspecting the order processing logs

7. The last line shows that the order was created. It's now time to look at the shipping service to see if it pulled the order created event from the RabbitMQ broker:

```
PS C:\> kubectl -n microserviceapp logs shippingprocessing-7bcfb96bfb-q6kbq shippingprocessing
←[40m←[32minfo←[39m←[22m←[49m: Microsoft.Hosting.Lifetime[0]
      Now listening on: http://[::]:80
←[40m←[32minfo←[39m←[22m←[49m: Microsoft.Hosting.Lifetime[0]
      Application started. Press Ctrl+C to shut down.
←[40m←[32minfo←[39m←[22m←[49m: Microsoft.Hosting.Lifetime[0]
      Hosting environment: Production
←[40m←[32minfo←[39m←[22m←[49m: Microsoft.Hosting.Lifetime[0]
      Content root path: /app
←[40m←[32minfo←[39m←[22m←[49m: TrackingService.Controllers.ShippingController[0]
      Received new event
←[40m←[32minfo←[39m←[22m←[49m: TrackingService.Controllers.ShippingController[0]
      4aadc0f8-eeda-4ee7-9c26-a6d39cbfbc28 (null) Created
←[40m←[32minfo←[39m←[22m←[49m: TrackingService.Controllers.ShippingController[0]
      Starting shipping process for order 4aadc0f8-eeda-4ee7-9c26-a6d39cbfbc28 with 2 products!
PS C:\> _
```

Figure 7.9 – Inspecting shipping logs

The last line also shows that the shipping process was started and that the order could be retrieved correctly from the order query service, because it found that two products were attached to it.

I hope that this small sample app can help you get started with microservices. Now, let's look at yet another trendy architecture style, namely serverless architecture.

Hands-on with a serverless architecture example

I introduced serverless architectures in *Chapter 6, Impact of the Cloud on the Software Architecture Practice*, so it is time to go through a small example to let you taste the serverless flavor. Remember that true serverless is based on fully delegating the infrastructure to the cloud provider, and costs are consumption-based. Therefore, it is hard to be cloud-agnostic, so I had to choose a cloud vendor for this example. Due to this, I went for Azure. The following diagram shows our very simple application:

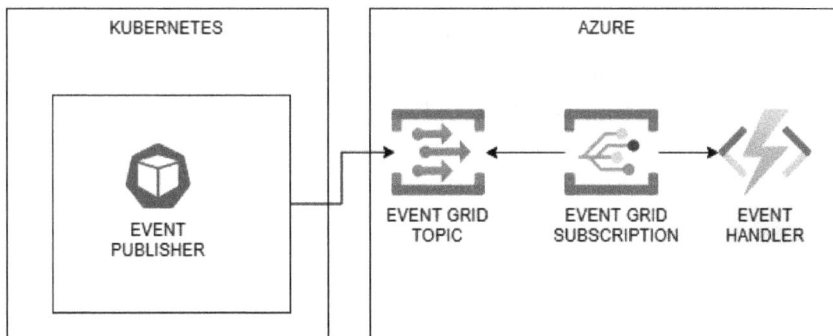

Figure 7.10 – Diagram of the serverless sample application

We are going to reuse the same K8s cluster to host the event publisher, which publishes events to a custom event grid topic. An Azure function, represented by the event handler in the preceding diagram, subscribed to our topic and gets triggered by the event grid whenever a new event is being pushed. As I explained earlier, serverless architectures are particularly suited for event-driven and asynchronous scenarios. The beauty of serverless is that you can immediately deploy your code and you do not especially need to worry too much about scalability, high availability, and so on. Since we're talking about code, let's see what the event publisher looks like.

Event publisher code

The event publisher is a simple console program that sends events to our event grid topic, which we will deploy in the next section. The program, shown in the following screenshot, takes the topic endpoint and the access key (to authenticate) as input, and then simply sends events every 100 milliseconds:

```
class Program{
    0 references
    static async Task Main(string[] args){
        EventGridPublisherClient client = new EventGridPublisherClient(
            new Uri(Environment.GetEnvironmentVariable("EvGridEndpoint")),
            new AzureKeyCredential(Environment.GetEnvironmentVariable("EvGridAccessKey")));
        while(true){
            await client.SendEventAsync(new EventGridEvent(
            "Serverless",
            "Serverless.OrderEvent",
            "1.0",
            Guid.NewGuid().ToString())
            );
            Thread.Sleep(100);
        }
    }
}
```

Figure 7.11 – Event publisher code

I could have used Dapr to push messages to the event grid, but the configuration of the Dapr component is not that straightforward, so I made it as simple as possible for you. I published this small console app on Docker Hub.

Deploying the required infrastructure

The time has come to deploy the Azure infrastructure and our event publisher within K8s. To deploy the Azure infrastructure, make sure to use an account that is at least a contributor to the subscription that will host the Azure resources. For your convenience, I have prepared a few IaC script templates, which are available in this book's GitHub repository. Let's go through the deployment steps.

1. First, you must log into `https://shell.azure.com/` using your trial or paid subscription.

2. Download the IaC files and the `.zip` package that contains the function code. Alternatively, you can clone the repository. Once downloaded, upload the files to Cloud Shell:

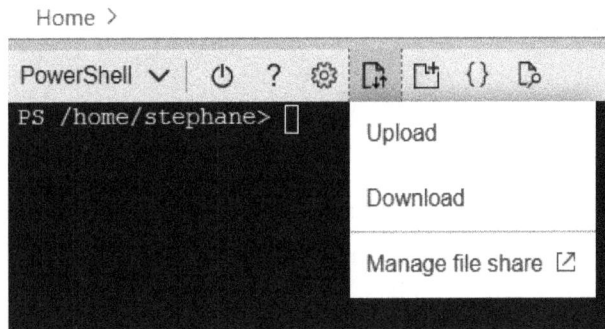

Figure 7.12 – Uploading files to Cloud Shell

3. You will have to upload files one by one.

4. Next, type the following command to create the resource group that will host the resources:

```
az group create -l westeurope -n packtserverless
```

5. Now, it is time to deploy the different services. Run the following command to deploy the first template:

```
az deployment group create --template-file serverless1.
json --resource-group packtserverless --parameters
appName="YOURVALUE"
```

6. Make sure you replace YOURVALUE with something unique. This will be used by Azure to connect to the function. In my case, I used `packtsrvlessarch`. Make sure not to use any fancy characters. At this stage, you should find the following resources in your resource group:

Name ↑↓	Type ↑↓	Location ↑↓	
📇 packtofsvc2yqt2cro	Storage account	West Europe	•••
⚡ packtsrvlessarch	Function App	West Europe	•••
💡 packtsrvlessarchofsvc2yqt2cro	Application Insights	West Europe	•••
📦 packtsrvlessarchofsvc2yqt2cro	App Service plan	West Europe	•••

Figure 7.13 – Checking the deployed resources

7. Azure functions require a **Storage Account** and an **App Service Plan** (bound to the dynamic pricing tier for serverless functions) to function. There is also an **Application Insights** instance that is used to monitor the function.

8. Now, you can deploy the function code that sits in the `.zip` file:

```
az webapp deployment source config-zip --resource-group
packtserverless --name YOURVALUE --src event-consumer.zip
```

9. Now comes the last part of the deployment, which is creating the event grid's topic and subscription. This last part will subscribe the previously deployed function to the topic:

```
az deployment group create --template-file
serverless2.json --resource-group packtserverless
--parameters eventGridTopicName='YOURVALUEtp'
eventGridSubscriptionName='YOURVALUEtpsub'
functionAppName='YOURVALUE'
```

10. Note that YOURVALUE still represents the name of your function. At this stage, you should have all resources deployed and linked together:

Name ↑↓	Type ↑↓	Location ↑↓	
packtofsvc2yqt2cro	Storage account	West Europe	•••
packtsrvlessarch	Function App	West Europe	•••
packtsrvlessarchofsvc2yqt2cro	Application Insights	West Europe	•••
packtsrvlessarchofsvc2yqt2cro	App Service plan	West Europe	•••
packtsrvlessarchtp	Event Grid Topic	West Europe	•••

Figure 7.14 – Checking if all the resources were deployed

11. The last item is the event grid topic. Upon clicking on it, you should see that a subscription for the Azure function was indeed created:

Name	Endpoint
packtsrvlessarchtpsub	AzureFunction

Figure 7.15 – The Azure function subscription

12. The name will be *YOURVALUE*tpsub. While you are looking at your event grid topic, take note of the topic's endpoint and access keys. The topic endpoint can be grabbed from the overview page, and the access keys (two keys) are accessible through the left menu. Copy only one of the keys. We need both the endpoint and one of the keys for our event publisher.

The Azure infrastructure has been completely deployed. However, we still need to deploy the event publisher to our K8s cluster before we can test the application. So, we are almost done. Let's go through the final remaining steps:

1. Edit `serverless.yml` (available on GitHub) and replace `YOURENDPOINT` and `YOURKEY` with your own values, taken from the preceding step:

```
spec:
  containers:
  - name: eventpublisher
    image: stephaneey/eventpublisher:dev
    env:
    - name: EvGridEndpoint
      value: "YOURENDPOINT"
    - name: EvGridAccessKey
      value: "YOURKEY"
    imagePullPolicy: Always
```

Figure 7.16 – YAML spec of the event publisher container

2. Deploy the YML file to your cluster:

```
kubectl apply -f .\serverless.yml
```

Congratulations, you are done!

Testing the application

Now that everything has been deployed, you should have at least one instance of the event publisher pod running. You can verify this as follows:

🖥 Windows PowerShell

```
PS C:\> kubectl get pod -n microserviceapp
NAME                                  READY    STATUS     RESTARTS    AGE
eventpublisher-68959df74-96w8x        1/1      Running    0           14s
PS C:\>
```

Figure 7.17 – Checking that the event publisher pod is running

This should already publish events to our grid, and notifications should be pushed to our function. Before we scale out to more instances, navigate to your Application Insights resource and click on the **live metrics** menu on the left. Once the metrics start showing up, scale out the event publisher, as shown in the following screenshot:

🖥 Windows PowerShell

```
PS C:\> kubectl scale deploy/eventpublisher --replicas=50 -n microserviceapp
deployment.apps/eventpublisher scaled
PS C:\> kubectl get pod -n microserviceapp
NAME                                  READY    STATUS              RESTARTS    AGE
eventpublisher-68959df74-25kfs        0/1      ContainerCreating   0           6s
eventpublisher-68959df74-294vj        0/1      Pending             0           6s
eventpublisher-68959df74-4v4ld        0/1      Pending             0           6s
eventpublisher-68959df74-5ph4z        0/1      Pending             0           6s
eventpublisher-68959df74-5szv6        0/1      ContainerCreating   0           6s
eventpublisher-68959df74-6qwd7        0/1      ContainerCreating   0           6s
eventpublisher-68959df74-72psp        0/1      Pending             0           6s
eventpublisher-68959df74-7hmvs        0/1      Pending             0           6s
eventpublisher-68959df74-7wjxl        0/1      Pending             0           6s
```

Figure 7.18 – Scaling out the event publisher

I scaled the event publisher for 50 instances. This may or may not work in your own environment, depending on your cluster capacity. If 50 is too much, just reduce it to 5. The goal is to create more events and see how Azure functions scale accordingly. The live metrics screen should show something similar to the following:

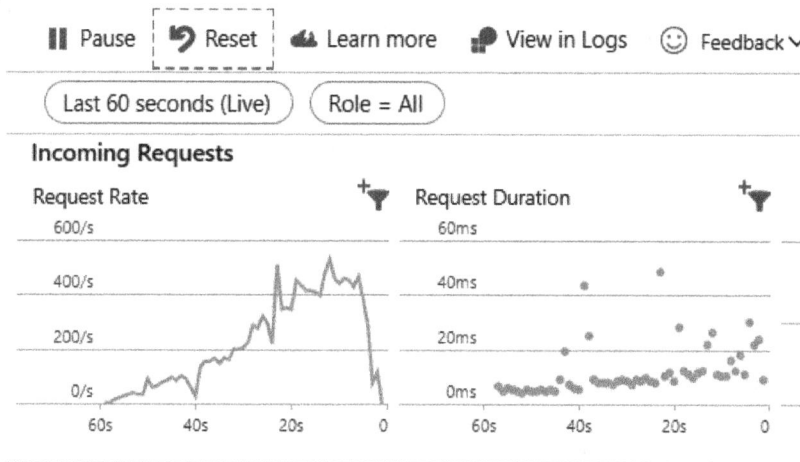

Figure 7.19 – Live metrics showing the request rate and duration

No matter how you scaled the event publisher, leave it running for about a minute and then scale the deployment back to zero.

As shown in the preceding screenshot, at peak time, the system was handling about 525 requests per second and most executions took less than 20 milliseconds. We did not have to configure anything; we just let the cloud provider adjust the computing power according to the demand. The following screenshot shows that the system scaled out, up to five instances:

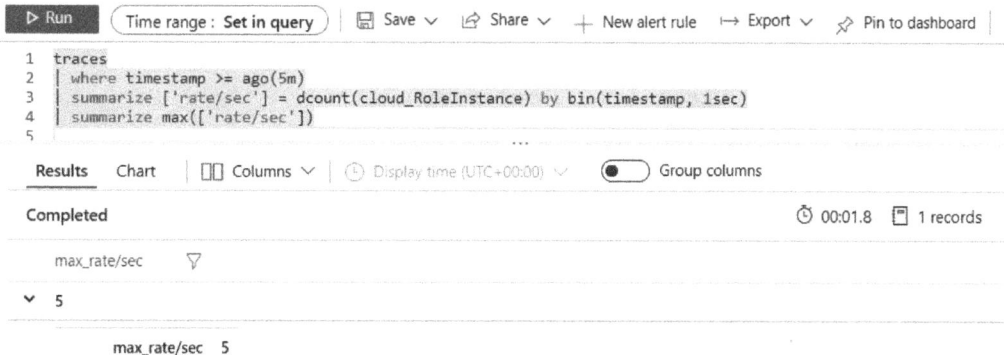

Figure 7.20 – Azure function max instance count

Running the same query after stopping the event publisher returns no results, meaning that an instance is no longer running. You do not need to run the query yourself. So, this very small serverless application is indeed based on a system that scales in and out according to demand. The cherry on the cake is that in Azure, you can have 1 million executions per month for free. You could force Azure functions to scale even more, should you have a powerful cluster or if you can run the event publisher from different systems at the same time. In my case, I simply used a single-node lab cluster, which is, by itself, a limiting factor, to really produce a high workload.

I hope that you enjoyed this short journey into the magical world of serverless architecture. Now, let's recap this chapter.

Summary

In this chapter, I tried to give you more concrete examples of trendy architecture styles, because I already introduced the theoretical part in the previous chapter. The purpose of the two examples provided in this chapter was to demonstrate how quickly you can get started with cloud and cloud-native applications. Both examples relied on Infrastructure as Code. The remaining manual steps were there to keep things simple but rest assured that this can be fully automated in the real world.

Both examples showed that the ecosystem plays an important role when building new solutions. For the serverless application, we relied on Azure Functions and Azure Event Grid, and we leveraged K8s's built-in scaling capabilities to load test our function handler. In the microservices example, we used Dapr, yet another great framework that comes from the K8s ecosystem. Both demos were intended to highlight the importance of this ecosystem, which you should never neglect as a software architect. The times where we were building everything from scratch are definitely gone. Studying and keeping in touch with the ever-growing ecosystem is entirely part of the software architect's duty.

Postface

I hope that you enjoyed your software architecture journey. As you understood from the initial chapter, there is no single vision of software architecture. I think that a good software architect must specialize in application development and architecture, as well as understand the bigger picture. A good software architect must be able to interact with every type of stakeholder. This is why I took you through the discovery of a few popular frameworks, such as TOGAF, ITIL, and NIST, as well as the ATAM methodology. These skills (even partially) are a must-have to grow as an architect. The frameworks help you structure and organize your work. The extent to which you apply them depends on the organization you are working for.

I could not bypass design patterns because they are an integral part of the software architect's job, but there are tons of books about them, so I did not want to write yet another book on design patterns. Our last two chapters showed how the cloud and related ecosystems are game-changers in terms of designing applications. This fact is often overlooked by many developers and architects, but do not fall into the same trap if you do not want to miss the bandwagon. It is clear that the cloud and what I would call the globalization of containerization are here to stay, so you'd better look at them. More importantly, I hope that you got the message I conveyed throughout the entire book: always exercise good judgment over things. Whether you are working with frameworks or design patterns, do not blindly apply things to the letter. Make sure you do not come with a one size fits all approach, which never works in practice, or at best generates much frustration. Be pragmatic, not dogmatic!

Packt>

Other Books You May Enjoy

If you enjoyed this book, you may be interested in these other books by Packt:

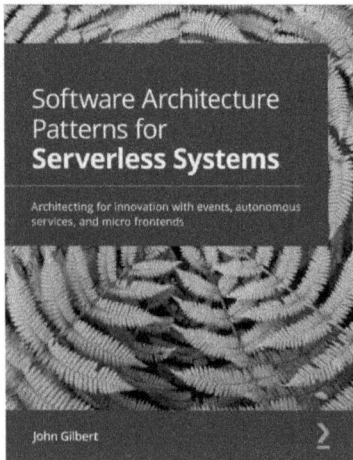

Software Architecture Patterns for Serverless Systems

John Gilbert

ISBN: 978-1-80020-703-5

- Explore architectural patterns to create anti-fragile systems that thrive with change
- Focus on DevOps practices that empower self-sufficient, full-stack teams
- Build enterprise-scale serverless systems
- Apply microservices principles to the frontend
- Discover how SOLID principles apply to software and database architecture
- Create event stream processors that power the event sourcing and CQRS pattern
- Deploy a multi-regional system, including regional health checks, latency-based routing, and replication
- Explore the Strangler pattern for migrating legacy systems

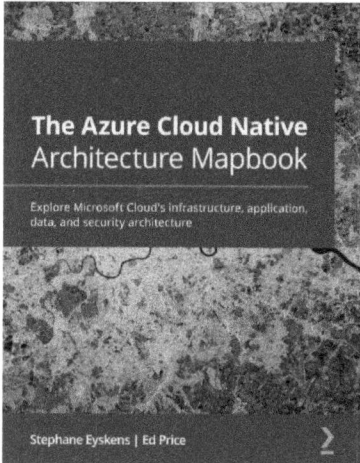

The Azure Cloud Native Architecture Mapbook

Stéphane Eyskens, Ed Price

ISBN: 978-1-80056-232-5

- Gain overarching architectural knowledge of the Microsoft Azure cloud platform
- Explore the possibilities of building a full Azure solution by considering different architectural perspectives
- Implement best practices for architecting and deploying Azure infrastructure
- Review different patterns for building a distributed application with ecosystem frameworks and solutions
- Get to grips with cloud-native concepts using containerized workloads
- Work with AKS (Azure Kubernetes Service) and use it with service mesh technologies to design a microservices hosting platform

Packt is searching for authors like you

If you're interested in becoming an author for Packt, please visit authors. packtpub.com and apply today. We have worked with thousands of developers and tech professionals, just like you, to help them share their insight with the global tech community. You can make a general application, apply for a specific hot topic that we are recruiting an author for, or submit your own idea.

Share Your Thoughts

Now you've finished *Software Architecture for Busy Developers*, we'd love to hear your thoughts! Scan the QR code below to go straight to the Amazon review page for this book and share your feedback or leave a review on the site that you purchased it from.

https://packt.link/r/1801071594

Your review is important to us and the tech community and will help us make sure we're delivering excellent quality content.

Index

T

U

W

www.ingramcontent.com/pod-product-compliance
Lightning Source LLC
Chambersburg PA
CBHW080552220326
41599CB00032B/6449